AL
in

for INFORMATION
TECHNOLOGY

By

Todd Verge

DELMAR
CENGAGE Learning

Australia • Brazil • Japan • Korea • Mexico • Singapore • Spain • United Kingdom • United States

DELMAR
CENGAGE Learning

**Practical Problems in Mathematics
for Information Technology
Todd Verge**

Vice President, Career and
Professional Editorial: **Dave Garza**

Director of Learning Solutions:
Sandy Clark

Executive Editor: **Stephen Helba**

Managing Editor: **Larry Main**

Senior Editorial Assistant:
Dawn Daugherty

Vice President, Career and
Professional Marketing:
Jennifer McAvey

Marketing Director: **Deborah S. Yarnell**

Marketing Manager: **Mark Pierro**

Marketing Coordinator: **Mark Pierro**

Production Director: **Wendy Troeger**

Production Manager: **Stacy Masucci**

Content Project Manager:
Christopher Chien

Art Director: **Benj Gleeksman**

Technology Project Manager:
Christopher Catalina

Production Technology Analyst:
Thomas Stover

For product information and technology assistance, contact us at
Professional & Career Group Customer Support, 1-800-648-7450

For permission to use material from this text or product, submit all requests
online at **cengage.com/permissions**
Further permissions questions can be e-mailed to
permissionrequest@cengage.com

Library of Congress Control Number: 2008920363

ISBN-13: 978-1-4283-2200-4

ISBN-10: 1-4283-2200-0

Delmar
5 Maxwell Drive
Clifton Park, NY 12065-2919
USA

Cengage Learning products are represented in Canada by Nelson Education, Ltd.

For your lifelong learning solutions, visit **delmar.cengage.com**

Visit our corporate website at **cengage.com**

Notice to the Reader
Publisher does not warrant or guarantee any of the products described herein or perform any independent analysis in connection with any of the product information contained herein. Publisher does not assume, and expressly disclaims, any obligation to obtain and include information other than that provided to it by the manufacturer. The reader is expressly warned to consider and adopt all safety precautions that might be indicated by the activities described herein and to avoid all potential hazards. By following the instructions contained herein, the reader willingly assumes all risks in connection with such instructions. The publisher makes no representations or warranties of any kind, including but not limited to, the warranties of fitness for particular purpose or merchantability, nor are any such representations implied with respect to the material set forth herein, and the publisher takes no responsibility with respect to such material. The publisher shall not be liable for any special, consequential, or exemplary damages resulting, in whole or part, from the readers' use of, or reliance upon, this material.

Printed in Canada
1 2 3 4 5 XX 10 09 08

This book is dedicated to my parents Joan and Ed for their unbounded selflessness.

Contents

PREFACE / ix

ACKNOWLEDGEMENTS / xi

APPENDIX / 125

GLOSSARY / 131

ANSWERS TO ODD-NUMBERED PROBLEMS / 135

Preface

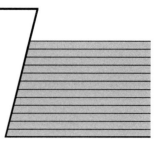

ABOUT THIS BOOK

Practical Problems in Mathematics for Information Technology is designed to provide students with the skills they will need to be successful in the field of information technology (IT). Each chapter is filled with practical problems taken directly from industry. These problems will not only reinforce the relevance of learning math but also introduce students to much of the terminology and abbreviations used in IT occupations today.

The text begins with review of basic mathematics and algebra and progresses through more advanced topics such as number systems, set theory, and logic. The book is designed to be used at either the secondary or postsecondary level. Instructors may choose to use this book as a stand-alone text or as a supplemental workbook to a theory-based text.

The Answers to Odd-Numbered Problems are provided at the end of this book along with a Glossary of technical terms and an Appendix containing symbols, tables, and other relevant information.

ABOUT THE AUTHOR

Todd Verge has a Bachelor of Science Degree in Physics from Dalhousie University, a Masters Degree in Philosophy from Carleton University, and a Diploma in Adult Education from the Nova Scotia Community College. He has worked for over 20 years as a technologist and a teacher. In 2006, Todd received the Nova Scotia Community College's Excellence in Teaching Award.

Acknowledgements

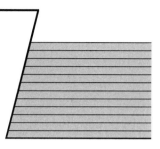

The author would like to thank his wife Janine and children, Katie and Ella, for their patience and sacrifice while this book was being written. He would also like to thank all the editors at Delmar Learning for their support and encouragement. Thanks go to Stephen Helba, Dawn Daugherty, and Christopher Chien. Your help was very much appreciated.

The author and Delmar Learning would like to acknowledge and thank the individuals who provided suggestions and comments during development of the manuscript. Thanks go to the review panel:

Robert Galante
Shawsheen Vocational Technical School
Billerica, MA

Anthony G. Russo
Cambridge R & L High School
Rindge School of Technical Arts
Cambridge, MA

Charulata Trivedi
Quinsigamond Community College
Worcester, MA

Harmit Kaur
Sinclair Community College
Dayton, OH

David Jellicoe
Nova Scotia Community College
Halifax, NS

Richard Goodpasture
Remington College
Jacksonville, FL

John Carpenter
Indiana Business College
Evansville, IN

Whole Numbers

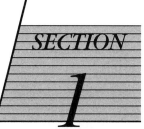 ## Unit 1 *ADDITION OF WHOLE NUMBERS*

BASIC PRINCIPLES

Whole numbers are the symbols used for counting amounts like 0, 1, or 2. They do not include decimal numbers like 0.5 or fractional numbers like 1/2. The placement of a whole number within a larger number determines its value. The digit on the right-hand side of a number is called the *least significant digit* because it has the least value. For example, the number 123 means 1 hundred, 2 tens, and 3 ones.

Two or more numbers can be added together to find a total or *sum* by placing them in a column and lining up the least significant digits. Starting with the right-hand column, the sum is found for each column. If the answer has two or more digits, only the least significant digit is written down while the rest is carried or added to the next column to the left.

Example: Find the sum of 15 + 72 + 133 + 256 + 91

$$
\begin{array}{r} {}^{1} \\ 15 \\ 72 \\ 133 \\ 256 \\ +\ 91 \\ \hline 7 \end{array}
\qquad
\begin{array}{r} {}^{21} \\ 15 \\ 72 \\ 133 \\ 256 \\ +\ 91 \\ \hline 67 \end{array}
\qquad
\begin{array}{r} {}^{021} \\ 15 \\ 72 \\ 133 \\ 256 \\ +\ 91 \\ \hline 567 \end{array}
$$

REVIEW PROBLEMS

1. A computer has two hard drives. One is 20 GB in size, and the other is 180 GB. What is the total storage size for both hard drives?   _____

2. A class "B" IP network uses 16 bits to specify the network address. If subnetting is used to add an additional 6 bits, what is the total number of bits in the extended network address?   _____

3. Your computer is running two user-mode processes. One is comprised of 15 threads, and the other has 29 threads. What is the total number of threads running from the two processes?   _____

4. A computer has one 512-MB and one 128-MB DIMM. What is the total amount of RAM in the computer?   _____

5. An IDE hard drive uses 38 watts of power, and the DVD drive uses 22 watts. How much power do they require together?   _____

6. ASCII represents 128 different characters and commands. If the extended ASCII set includes an additional 128 characters, what is the total number of characters and commands in the extended ASCII set?   _____

7. A Linux computer is running 19 foreground processes and 52 background processes. How many processes are running in total?   _____

8. You need to purchase some 1-GB DIMMs to upgrade the memory in one lab that contains 40 computers and in another lab with 55 computers. How many DIMMs will you need to purchase?   _____

9. A network administrator is responsible for the computers in three offices. One office has 15 computers, another has 47, and the last has 11. How many computers is she responsible for?   _____

10. A network patch cable (from the computer to the wall) is 14 feet long. A second cable runs 52 feet to a punch down block in the server room, and a 4-foot cable connects the punch down block to a switch. What is the total length of the cable run from the computer to the switch?   _____

11. You need to purchase a site license for a word-processing program. Your organization has 62 computers in the finance department, 55 in marketing, 15 in administration, and 521 on the factory floor. How many computers must your site license support?   _____

12. You need to save a backup copy of five different files. They are 64 KB, 1024 KB, 512 KB, 64 KB, and 128 KB in size. How much room will it take to store all five files?   _____

13. You work in a technical support call center and keep track of your daily calls in the chart below. What is your total number of calls for the week? _____

Day	No. of Calls
Mon	52
Tue	180
Wed	107
Thu	94
Fri	44

14. A network switch connects LAN segments. If the four segments attached to a switch have 32, 63, 27, and 6 computers attached, how many computers are on the network? _____

15. You are upgrading a computer. If it typically takes you 45 minutes to install a new motherboard, 15 minutes to install a hard drive, and 60 minutes to upgrade an operating system, how long do you estimate the whole job will take? _____

16. A network administrator is taking an inventory of user accounts in the system. She determines that there are 1,237 active accounts and another 415 disabled ones. What is the sum of all the accounts in the system? _____

17. An incremental backup is saved to tape every weeknight. The size of each backup is listed in the following table. What is the sum of all the backups? _____

Day	Size of Backup
Mon	1,247 MB
Tue	382 MB
Wed	6,640 MB
Thu	15 MB
Fri	221 MB

18. You need to decide if it is less expensive to upgrade your computer to support a new operating system or purchase a new system. The required new parts for the upgrade are: motherboard at $82, CPU at $150, RAM at $65, and video card at $115. A new computer costs $399. Which is less expensive? _____

19. You are making network cables. You need to make six cables with lengths of 150 feet, 55 feet, 350 feet, 15 feet, 310 feet, and 90 feet. You have a 1,000-foot spool of cable. Do you have enough cable to finish the job? _____

20. You are assembling a computer from the parts in the following table. What is the total cost of the computer? _____

Part	Cost
Case	$48
Motherboard	$95
CPU and Fan	$317
Hard Drive	$180
Video Card	$110
Network Card	$26
Monitor	$150
DVD-ROM	$45

Unit 2 SUBTRACTION OF WHOLE NUMBERS

BASIC PRINCIPLES

Subtraction is used to find the difference between two numbers. This difference is called the remainder. Just as in addition, the numbers are put into columns with the least significant digits aligned.

Example: Subtract 123 from 759.

$$\begin{array}{r} 759 \\ -\ 432 \\ \hline 327 \end{array}$$

Example: Subtract 273 from 392.

$$\begin{array}{r} 3\overset{8}{\cancel{9}}2 \\ -\ 273 \\ \hline 119 \end{array}$$

In this case, the number that was first being subtracted (3) was larger than the number it was being subtracted from (2). To avoid a negative result, one digit was *borrowed* from the next column on the left to make the remainder $12 - 3 = 9$. In other words, to make subtraction easier, 392 was changed from $(3 \times 100) + (9 \times 10) + (2 \times 1)$ to $(3 \times 100) + (8 \times 10) + (12 \times 1)$.

REVIEW PROBLEMS

1. You work at a computer store that charges $113 to upgrade the RAM in a computer but offers to buy back the old RAM for $25. What does the upgrade cost? _____

2. A MAC address is 48 bits long. If the first 24 bits are used to identify the manufacturer, how many bits are left to be assigned by the vendor? _____

3. One database field shows the start time as 5:17 (hours and minutes), and another field shows the end time as 5:56. How much time elapsed between the start and end times? _____

4. A programmer is debugging 1,465 lines of source code. If 519 of the lines are comments, how many lines are left to be compiled into an exactable program? _____

5. The cost of a hard drive drops from $117 to $89. How much are you saving?   _____

6. A class "A" IP network uses 24 bits to specify the host address. If subnetting is used to borrow 9 bits, how many bits are left in the host address?   _____

7. A computer with 512 MB of RAM has an onboard video card that borrows its memory from the motherboard. How much system memory is left if it borrows 128 MB of RAM?   _____

8. A blank CD-R will hold 700 MB of data. How much space is left after you write a 257 MB video file and a 14-MB audio file?   _____

9. A computer shop sold 176 external hard drives before Christmas but then refunded the money for 18 of them that were returned on December 26. How many were not returned?   _____

10. A network installer buys a 1,000-foot spool of Cat 5e cable. She uses 344 feet on the first day and 539 feet on the second day of the job. How much cable is left for the third day?   _____

11. Your computer takes 61 seconds to boot. You can save 5 seconds by removing unwanted services, 2 seconds by disabling some unused hardware, and 4 seconds by changing the boot order of your drives. How long will it take your computer to boot if you make all of these changes?   _____

12. A senior programmer makes $2,745 every two weeks. Deducted from this is $644 for taxes, $245 for pension, $65 for union dues, and $361 for insurance. How much is left over?   _____

13. The symbol "1" is represented by the ASCII value 49. If each value is 48 more than the number it represents, what symbol is represented by the ASCII value 53?   _____

14. A Web designer has divided a page into three frames. The first frame has a width of 45 pixels, and the second has 160 pixels. If the whole page is 640 pixels wide, how wide is the third frame?   _____

15. The maximum allowable cable distance from a computer to a network switch is 328 feet. If a 4-foot cable connects the switch to a punch down block in the server room, and a 297-foot cable extends from the punch down block to an RJ-45 jack in an office, what is the maximum cable length left to connect a computer to this jack?   _____

16. You have a 120-GB hard drive with a 60-GB primary partition and a 25-GB extended partition. How much unpartitioned space is left on the drive? _____

17. A 24-port switch is currently connected to 16 Cat 6 cables. If 7 more cables are attached, how many free ports are left? _____

18. A computer has a 300-watt power supply. The peak power requirements for all of its components are listed in the chart below. Is there enough power available to add a second 41-watt hard drive without upgrading the power supply? _____

Component	Power Required
Motherboard	18 watts
CPU and Fan	125 watts
RAM	80 watts
DVD Burner	22 watts
Hard Drive	41 watts

19. Your 160-GB hard drive is completely full. If you remove 10 GB of programs, 23 GB of video files, and 15 GB of audio files, will the used space be below 100 GB? _____

20. You have $1,400 to purchase a computer system. If you purchase the parts below, how much money do you have left for software? _____

Part	Cost
Case	$67
Motherboard	$107
CPU and Fan	$240
Hard Drive	$233
Video Card	$89
Monitor	$199
DVD-ROM	$48

Unit 3 MULTIPLICATION OF WHOLE NUMBERS

BASIC PRINCIPLES

Multiplication is a specialized form of addition. Suppose you wanted to add together five groups of three objects. You could add 3 to itself five times, or you could simply multiply 3 by 5. In both cases the answer or *product* is 15 (3×5 is the same as $3 + 3 + 3 + 3 + 3$).

```
    3
    3
    3
    3          5
  + 3        × 3
   15         15
```

Just like in addition, the numbers to be multiplied are placed in a vertical column with the least significant digits aligned. Multiplication also begins on the right-hand side, and numbers are carried when necessary. **Hint:** It's often easier to put the larger number on top.

Example: Find the product of 48 and 2.

```
   1           1
  48          48
 × 2         × 2
   6          96
```

The solution is found by starting on the right and multiplying 2 by 8. The answer, 16, is broken into two parts. The 6 is written below the right-hand column, and the 1 (from the tens column) is carried to the next column on the left so that it can be added later. Next the 2 is multiplied by 4 and then added to the 1 that was just carried from the previous step for a total of 9. The product of 48 and 2 is 96.

Example: Find the product of 317 and 5.

```
    3         0 3       1 0 3      1 0 3
  317        317        317        317
 ×  5       ×  5       ×  5       ×  5
    5         85        585       1585
```

The solution is found by first multiplying 5 by 7 ($5 \times 7 = 35$). The least significant digit in the answer is 5, so it is written below that column while the 3 from the tens column is carried. The next digit to the left, which is 1, is multiplied by 5 ($5 \times 1 = 5$) and then added to the number that was carried ($5 + 3 = 8$). This answer is then written in the tens column below. Finally, the third digit is multiplied by 5 ($5 \times 3 = 15$), and this value is also written below, in the hundreds column.

If both of the numbers you are multiplying have more than one digit then you have to solve a series of multiplications and add their products.

Example: Find the product of 12 and 36.

```
  1          1          1          1          1
  36         36         36         36         36
× 12       × 12       × 12       × 12       × 12
  2          72         72         72         72
                        60         360      + 360
                                              432
```

Because 36 is simply the sum of 30 and 6, the product of 12 and 36 can be found by adding the product of 6 times 12 ($6 \times 12 = 72$) and the product of 30 times 12 ($30 \times 12 = 360$). The sum is 432 ($72 + 360 = 432$).

Example: Find the product of 123 and 48.

```
                                      1          1          1
  2        1 2        1 2        1 2        1 2        1 2
  123        123        123        123        123        123
×  48      ×  48      ×  48      ×  48      ×  48      ×  48
  4          84         984        984        984        984
                                   20         920      +4,920
                                                        5,904
```

Because 48 is simply 4 tens plus 8 ones or $(4 \times 10) + (8 \times 1) = (40 + 8) = 48$, the answer is the sum of two different products: $(8 \times 123) + (40 \times 123) = (984 + 4,920) = 5,904$.

REVIEW PROBLEMS

1. A computer contains three 256-MB memory modules. How much memory does it have?   _____

2. If a USB 2.0 hard drive can transfer 60 MB per second, how many megabytes can it transfer in one minute?   _____

3. A motherboard has a 533-MHz frontside bus and runs at a 4× multiplier. Find the product to determine the clock speed of the CPU.   _____

4. A PCI bus has a 4-byte data path and a bus speed of 66 MHz. Multiply these numbers to determine the maximum throughput in MB/s.   _____

5. A screen has a resolution of 640 × 480 pixels. Find the product to determine the number of pixels on the display?   _____

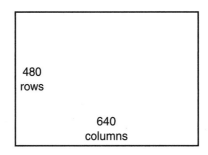

6. If a screen has a resolution of 1,920 × 1,200 and a color depth of 32 bits or 4 bytes per pixel, how much memory is used to display the screen?   _____

7. There are 8 bits in 1 byte. How many bits are in a 128-byte text file?   _____

8. A 20-GB hard drive is formatted with FAT32. If each sector is 512 bytes, and there are 32 sectors per cluster, then how many bytes are in each cluster?   _____

9. A computer power supply draws 3 A and 120 V. If watts equal volts multiplied by amperes, how many watts is the power supply using?   _____

10. USB 1.1 transfers data at 12 Mb/s, and USB 2.0 is 40 times faster. How fast is USB 2.0?   _____

11. A two-minute video file contains 30 frames per second. How many frames are in the video?   _____

12. A 1× CD transfers data at 150 KB/s. How fast does a 52× CD transfer data?   _____

13. If you earn $15 per hour entering data, how much will you earn in a 45-hour week?   _____

14. CD-quality audio takes 44,100 samples per second at 2 bytes per sample. If two channels are recorded (stereo) for one minute, how large is the uncompressed .wav file?   _____

15. If a printer produces 150 dpi (dots per linear inch), how many dots are in a 4-inch × 6-inch picture?   _____

16. There are 1,024 bytes in one KB. How many bytes are in 512 KB?   _____

17. PRI (ISDN) consists of twenty-three 64-Kbps bearer channels. How fast is PRI?   _____

18. An IP address is made up of 4 × 8-bit octets. How many bits are in an IP address?   _____

19. A MAC address is made up of 6 × 8-bit hexadecimal numbers. How many bits are in a MAC address?   _____

20. A network backup program can copy 35 MB/s. How many MB can it back up in an hour?   _____

▢ Unit 4 DIVISION OF WHOLE NUMBERS

BASIC PRINCIPLES

Division is a specialized form of subtraction where a smaller number is subtracted from a larger one a certain number of times. The larger number is called the *dividend* and the smaller is the *divisor*. The number of times the divisor can be subtracted from the dividend is called the *quotient*.

A division problem may be represented in two different forms.

$$\text{Dividend} \div \text{Divisor} = \text{Quotient} \quad \text{- or -} \quad \text{Divisor} \overline{)\text{Dividend}}^{\text{Quotient}}$$

Example: Find the quotient of 69 ÷ 3.

$$
3\overline{)69}
\qquad
\begin{array}{r}
2 \\
3\overline{)69} \\
-60 \\
\hline
09
\end{array}
\qquad
\begin{array}{r}
23 \\
3\overline{)69} \\
-60 \\
\hline
9 \\
-9 \\
\hline
0
\end{array}
$$

The solution is found by starting with the divisor (3) and dividing it into the first digit on the left of the dividend (6). The answer (2) is then written above the overbar. The divisor (3) is then divided into the next digit to the right in the dividend (9), which gives an answer (3) that is also written above the overbar. The quotient is 23.

Example: Find the quotient of 645 ÷ 15.

$$
\begin{array}{r}
0\mathbf{4} \\
15\overline{)645} \\
-60 \\
\hline
\mathbf{4}
\end{array}
\qquad
\begin{array}{r}
4 \\
15\overline{)645} \\
-60 \\
\hline
45
\end{array}
\qquad
\begin{array}{r}
43 \\
15\overline{)645} \\
-60 \\
\hline
45 \\
-45
\end{array}
\qquad
\begin{array}{r}
43 \\
15\overline{)645} \\
-60 \\
\hline
45 \\
-45 \\
\hline
\mathbf{0}
\end{array}
$$

In this case, start with the first digit of the dividend and determine how many times the divisor can be subtracted. The answer is 0 because 15 can't be subtracted from 6. Similar to the way you carry numbers in addition and multiplication, you can carry the number that isn't subtractable and use it with the next digit to the right in the dividend (64). Now ask how many times 15 can be subtracted from 64. The answer is written on the overbar above the appropriate digit, becoming the first value in the quotient. The remainder (4) is written below. (Keep in mind that the 4 is in the tens column, which means that it represents 40.) Now combine the 4 with the last digit in the dividend to give 45. How many times will 15 divide into 45? The answer is exactly 3 times, so there is no remainder. We will deal with leftover remainders in the unit on decimal fractions.

REVIEW PROBLEMS

1. An 802.11-n wireless network has twice the maximum bandwidth of 802.11 g. If 802.11 n broadcasts at 108 Mbps, what is the maximum bandwidth of an 802.11-g wireless network? _____

2. A company spends $39,000 on a site license for a word-processing program. If the company has 750 computers, how much is it spending per computer? _____

3. A networking job requires 350 RJ-45 connectors. If the connectors come in boxes of 25, how many boxes are needed? _____

4. CPUs typically run at a multiple faster than the system bus on the motherboard. If a Pentium IV has a clock speed of 2,000 MHz, and the system bus runs at 400 MHz, what is the multiplier? _____

5. The 384 KB of upper memory on a Windows 9x computer is divided into three equal sections. How large is each section? _____

6. A digital photograph is compressed 20 to 1 before being attached to an e-mail. How small would 1,360 KB become after maximum compression? _____

7. Your antivirus program can scan 287 files per second. If your computer contains 13,776 files, how long will the scan take? _____

8. If a programmer can debug 120 lines of code in an hour, how many hours will it take to debug 6,600 lines of code? _____

9. A T1 Internet connection provides maximum download speeds of 1,536 Kbps. If 128 computers are sharing this connection, what is the average download speed per computer? _____

10. If a computer is running at 3,000 MHz (3,000 million cycles per second) and it takes 40 cycles to do each operation, how many operations can it do per second? **Note:** Hz (hertz) stands for cycles per second. _____

11. The default typematic rate for a computer keyboard is six characters per second. How long would you have to keep pressing a key to repeat a character 150 times? _____

12. How many times faster is a (4,000 MB/s) 16× PCI Express expansion bus than an older (16 MB/s) ISA bus? _____

13. You are taking an on-line accreditation test that contains 120 questions. Will you be able to finish it in 35 minutes if you can answer an average of 3 questions per minute? _____

14. If a virus is infecting an average of six computers per hour. How long will it take to infect a lab with 48 computers? _____

15. Partial T1 connections come in 64 Kbps blocks. How many would be required to create a 768 Kbps connection? _____

16. A 228-foot cable run is bundled with cable ties every three feet. How many cable ties are used? _____

17. A building with 15 offices has 420 network connections. If each office has the same number of connections, how many are there per office? _____

18. Many hard drives spin at 7,200 rpm (rotations per minute). How many rotations do they spin per second? _____

19. You have a 380-GB hard drive and you want to partition it into four equal parts. How large should you make the partitions? _____

20. There are 1,024 bytes in a kilobyte. How many kilobytes are in 15,360 bytes? _____

Unit 5 COMBINED OPERATIONS WITH WHOLE NUMBERS

BASIC PRINCIPLES

In the real world, you often have to solve problems that involve combining the operations of addition, subtraction, multiplication, and division. When presented with such complicated problems, there is a particular order in which you must deal with combined operations.

1. Do operations that are inside parentheses.

2. Solve expressions that contain exponents or roots.

3. Multiply or divide from left to right.

4. Add or subtract from left to right.

Many people use the acronym BEDMAS to help remember the proper order of operations. It stands for **B**rackets, **E**xponentials, **D**ivision and **M**ultiplication, **A**ddition, and **S**ubtraction.

Example: Solve the equation $(5 + 3) + 6^2 \times (7 - 2)$.

$$(5 + 3) + 6^2 \times (7 - 2) = 8 + 6^2 \times 5 = 8 + 36 \times 5 = 8 + 180 = 188$$

REVIEW PROBLEMS

1. A database has six fields that vary in length according to the table below.
    If there are 417 entries in each field how large is the database?   _____

Field Name	Size
First_Name	24 bytes
Last_Name	36 bytes
Address_1	128 bytes
Address_2	128 bytes
State	18 bytes
Zip	5 bytes

2. Assembling a computer takes a computer technician 15 minutes. How many computers can be assembled by three technicians in five 8-hour days? _____

3. One minute of uncompressed CD-quality audio takes up 10 MB of storage space. If MP3 files have a 12:1 compression ratio, how many minutes of music from an MP3 can fit onto a 700-MB CD-R? _____

4. If RAM costs $85 per GB, how much will it cost to upgrade a lab of 35 computers from 1 GB to 4 GB of RAM? _____

5. If a video card can display 2,048 × 1,536 (3 megapixels) in 24-bit color (3 bytes per pixel) at 85 Hz (cycles per second), what size frame buffer would it take to store two seconds of video? _____

6. A Web designer took 15 hours to design a 21-page Web site. Should she charge $195 an hour or $145 a page? _____

7. An external SATA interface is five times faster than USB 2.0, and USB 2.0 is 40 times faster than USB 1.1. If USB 1.1 transfers data at 12 Mbps, how fast is external SATA? _____

8. There are 8 bits in a byte, 1,024 bytes in a kilobyte, and 1,024 kilobytes in a megabyte. How many bits are in one megabyte? _____

9. A video game designer is adding texture to the walls of a virtual environment. There are 4 walls per room, 12 rooms per building, and 3 buildings. If each wall takes five minutes to texture, how many hours will the job take? _____

10. An antivirus program can scan 15 executable files per second and 220 data files per second. If the computer contains 330 executable files and 3,080 data files, how long will the scan take? _____

11. If RJ-45 connectors cost 35 cents each, and Cat 5 cable costs 12 cents per foot, what would it cost to make twenty-four 3-foot cables? **Note:** Cables need an RJ-45 connector at each end. _____

12. Your computer has 512 MB of RAM but reserves 64 MB for an onboard video card, leaving 448 MB of available RAM. How much available RAM would you have if you doubled the system RAM and doubled the amount reserved for the video card?

_____

13. High memory on a Windows 9$\underline{x}$ machine is 64 KB – 16 B. What size is the high memory area in bytes? **Note:** 1 KB = 1,024 B.

_____

14. A 3,000-MHz computer normally runs at 55 degrees Celsius. The CPU runs 1 degree hotter for every additional 7 MHz. How fast can it be overclocked before it reaches a temperature of 72 degrees Celsius?

_____

15. An electrical signal travels 30 cm (centimeters) in one nanosecond. If a signal takes 12 nanoseconds to reach the end of a cable and reflect back, how long is the cable?

_____

16. A programmer is normally paid $38 per hour but gets an overtime rate of $57 per hour for any weekly hours over 40. Given the time sheet below, how much will this programmer make?

_____

Day	Hours Worked
Monday	8
Tuesday	12
Wednesday	15
Thursday	5
Friday	16
Saturday	7
Sunday	0

17. A help-desk technician receives 8 calls an hour. During an 8-hour day, a service technician had to be sent out 5 times for problems that couldn't be solved over the phone. How many problems were successfully solved over the phone?

_____

18. An old 3.5-inch floppy disk reads from both sides. Each side has 80 tracks, 18 sectors per track, and 512 bytes per sector. How many bytes of data will one disk hold?

_____

19. An old 3.5-inch floppy disk reads from both sides at once. Each side has 18 sectors per track (once around the disk) and 512 bytes per sector. If the disk spins at 5 rotations per second, what is the data transfer rate in bytes per second?   _____

20. Each month you need to make one 27-MB full backup, three 12-MB differential backups, and twenty-four 4-MB incremental backups. Will a 160-MB tape hold all the backups?   _____

Common Fractions

Unit 6 ADDITION OF COMMON FRACTIONS

BASIC PRINCIPLES

When a whole number is divided by a larger whole number, the result will be too small to be a whole number itself. This partial number is called a *common fraction*. For example, when 2 is divided by 3, the result (being less than 1) is expressed as $\frac{2}{3}$ or 2/3.

There are two parts to a common fraction. The number above the line is called the *numerator* and the number below is called the *denominator*. Suppose you cut a pie into 4 equal pieces and then take 3 for yourself. The total number of pieces (4) is the denominator, and the number you took (3) is the numerator. By placing the numerator over the denominator you can see that you took 3/4 of the pie!

Common fractions can only be added when the denominators are all the same (that means that they have a *common denominator*). Because there is a common denominator, only the numerators must be added, and their sum is placed over the common denominator.

Example: Find the sum of 2/7 + 3/7.

The numerators 2 and 3 add to 5. Placing this number over the denominator gives a sum of 5/7.

FINDING A COMMON DENOMINATOR

Often the fractions you want to add do not have the same denominator. In that case, you must first find a common denominator—a number that all the denominators can divide into evenly (with no remainder).

Example: Find the sum of 2/5 + 1/3.

2/5 + 1/3 = 6/15 + 5/15 = 11/15

A common denominator of 2/5 and 1/3 is 15 because both 5 and 3 will divide into it evenly. The original fractions (2/5 and 1/3) must now be changed into *equivalent fractions* with this common denominator. With 2/5, this is accomplished by dividing the original denominator (5) into the common denominator (15), which gives a quotient of 3. The quotient (the answer) must now be multiplied by the original numerator (2), for a product of 6. Now you have a new fraction of 6/15 which is equivalent to the original fraction of 2/5. Use the same procedure to find an equivalent fraction of 1/3 with a denominator of 15. When the numerators are then added together, their total is written over the common denominator as 11/15.

REDUCING FRACTIONS TO THE LOWEST TERMS

In order to keep problems as simple as possible, equivalent fractions should always be expressed in the lowest possible terms. In other words, the *lowest* common denominator should be found. This is done by dividing the numerator and denominator by the largest number that does not leave a remainder.

Example: Reduce the fraction 3/6 to the lowest terms.

$$\frac{3}{6} = \frac{3(\div 3)}{6(\div 3)} = \frac{1}{2}$$

Another type of fraction reduction is needed when the numerator is larger than the denominator (3/2), which is called an *improper fraction.* An improper fraction can be reduced by dividing the numerator by the denominator, which results in a whole number with a remainder. The remainder is placed over the denominator to make a new fraction. Because the answer is now comprised of a whole number and a fraction, it is called a *mixed number.*

Example: Express the improper fraction 7/3 as a mixed number.

$$\frac{7}{3} = 2 \text{ (remainder of 1)} = 2\frac{1}{3}$$

Example: Find the sum of 3/5 + 2/3 + 1/4.

$$\frac{3}{5} + \frac{2}{3} + \frac{1}{4} = \frac{36 + 40 + 15}{60} = \frac{91}{60} = 1\frac{31}{60}$$

The first step to a solution is finding a common denominator of 3/5, 2/3, and 1/4. The easiest way to find a common denominator is to multiply all of the original denominators together. The product of $5 \times 3 \times 4 = 60$, so we know that 60 is a common denominator. Equivalent fractions are found with the common denominator and their numerators are added together.

REVIEW PROBLEMS

1. A USB cable has two sections. One is 3 1/2 feet long, and the other is 6 3/4 feet. How long is the overall cable? _____

2. A CD cleaning solution is made by mixing 2 1/2 liters of distilled water with 1/3 liter of concentrated cleaner. How much solution is made? _____

3. The DC power connector in a computer is already 5 7/8 inches long. If a 3 7/16-inch extension is added, how far can the power connecter now reach? _____

4. A video signal travels 3 1/2 feet to a KVM and another 22 3/4 feet to an LCD projector in the ceiling. How far does the video signal travel? _____

5. There are several boxes of 120 RJ-45 connectors left over from a networking job. One box is 1/2 full, one is 3/4 full, and one is only 1/3 full. How many connectors do you have in total? _____

6. A computer is equipped with an SCSI-3 interface that can connect up to 15 devices. If 1/5 of the devices being used are hard drives and another 1/3 of the devices are external peripherals, how many SCSI devices does the computer have? _____

7. The TTL (time to live) for a packet is set at 24 hops. If a packet uses up 1/4 of its hops getting to router "A" and another 1/3 of its hops to get to router "B," how many hops has it used? _____

8. Find the shortest length of networking cable from which the following sections can be cut: 3 1/2 feet, 12 3/4 feet, 17 1/3 feet, and 18 1/4 feet. _____

9. A coaxial cable is connecting several workstations in a linear bus network. The distance between the computers is 5 1/5 m, 9 3/10 m, 25 1/2 m, and 18 9/10 m. What is the total length of the cable connecting the computers? _____

10. A desktop publisher is creating a page layout that has two 3 3/4-inch columns separated by a 1/4-inch space. How wide is the total layout area? _____

11. A programmer is debugging code. On the first day, he is able to debug 1/5 of the code; on the second day, he only gets through 1/8; and on the third day he debugs 1/4 of the code. How much of the code has been debugged? _____

12. A computer contains three memory modules (of RAM) of the following sizes: 1/2 GB, 2 GB, and 1/4 GB. How much RAM is in the computer?   _____

13. This motherboard attaches to the case with metal brackets attached to each of the holes. How wide apart are the two outermost holes?   _____

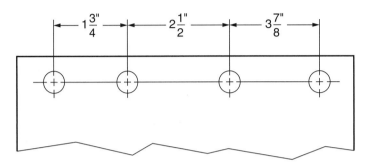

14. A large organization uses subnetting with their network. If 1/2 of the available subnets is used by the head office, 1/5 by the sales department, and 1/4 by marketing, what fraction of the available subnets is being used?   _____

15. You wish to attach this fan to your computer's enclosure to prevent overheating and increase air flow. How far apart do the drill holes for the mounting bolts need to be?   _____

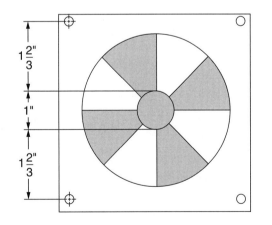

16. The chart below shows the time it takes to install each of the components of a computer. How long does the whole process take? _____

Component	Time Needed
Motherboard	5½ minutes
Processor	1¾ minutes
RAM	¼ minute
Hard Drive	1¼ minutes
Optical Drive	½ minute

17. This bolt is used to attach a motherboard to the case. How far is the top of the bolt from the bottom of the case? _____

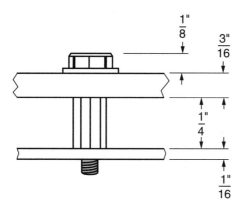

18. A technician spent 1/4 of an hour assembling a computer, 1 1/2 hours installing the operating system, 3/4 of an hour installing drivers and updates, 1 1/3 hours installing applications, and 4 hours for burn-in testing. How long did the whole process take? _____

19. A programmer keeps track of working hours on a time card. How many hours did he work this week?   _____

Day	Hours Worked
Monday	8½ hours
Tuesday	6¾ hours
Wednesday	7¾ hours
Thursday	5¼ hours
Friday	6½ hours
Saturday	6¾ hours
Sunday	Day Off

20. You are making network cables. You need to make six cables with lengths of 75 1/2 feet, 56 3/4 feet, 155 feet, 15 3/4 feet, 110 1/2 feet, and 75 1/4 feet. You have a 500-foot spool of cable. Do you have enough cable to finish the job?   _____

 *Unit 7* **SUBTRACTION OF COMMON FRACTIONS**

BASIC PRINCIPLES

Just as it is in adding fractions, subtracting fractions can only take place if they have a common denominator. Once a common denominator has been found, the numerators can then be subtracted, and the resulting fraction can be expressed in its lowest terms.

Example: Subtract 1/3 from 3/4.

$$\frac{3}{4} - \frac{1}{3} = \frac{9}{12} - \frac{4}{12} = \frac{5}{12}$$

Example: Subtract 1/2 from 6/8.

$$\frac{6}{8} - \frac{1}{2} = \frac{12}{16} - \frac{8}{16} = \frac{4}{16} = \frac{1}{4}$$

REVIEW PROBLEMS

1. You have two partially full ink cartridges for your printer. The first is only 1/3 full, while the second is still 3/4 full. How much more ink is in the second cartridge? _____

2. Of the symbols used in the hexadecimal numbering system (base 16), 5/8 of the symbols are numbers and the rest are letters. What fraction is letters? _____

3. You are using PoE (Power over Ethernet) to power a remote repeater. You attach an 8-volt power supply, which reduces the voltage over the cable run by one-third. If the repeater requires at least 5 volts, will it have enough power? _____

4. If spam accounts for 3/5 of the e-mail messages sent, what fraction of e-mail is not spam? _____

5. You use the Ping command to test a network connection. If 3/4 of the packets sent are lost, what fraction reaches their destination? _____

6. An IP address is 32 bits long. If 3/16 of the bits are used for the host address, how many bits are left for the network address?   _____

7. The drive cable below will connect to two hard drives. How far can the first drive be from the motherboard?   _____

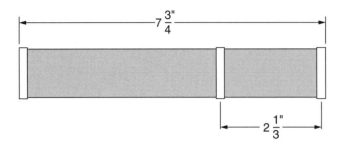

8. A network technician has a 100-foot spool of Cat 5 cable. She uses 12 1/2 feet on one job, and 53 3/4 feet on another job. How much cable is left on the spool?   _____

9. Your company does three kinds of backups; full, differential, and incremental. If 1/30 of the backups are full, and 1/10 of the backups are differential, what fraction is incremental?   _____

10. In an 80-conductor IDE cable, 1/2 of the wires are used as ground to reduce cross talk, and 1/5 carry data. What fraction is used for other purposes?   _____

11. An office has two computers that connect to the same wall plate. If the first computer is 15 3/4 feet from the wall plate, and the second computer is 6 1/2 feet in the same direction, how far apart are the computers?   _____

12. A Web page is 800 pixels wide. If a table starts halfway across the page and extends to three-quarters of the way across, how wide is the table?   _____

13. The protective jacket on an optical fiber has an outside diameter of 1/4 inch and is 3/32 of an inch thick. What is the inside diameter of the jacket?   _____

14. A computer is connected to a 9 1/2-foot power cable. If the cable is replaced with one that is only 5 3/4 feet long, how much cable length has been lost?   _____

15. In an ASCII table, 1/4 of the symbols are control characters, 13/32 of them represent letters, and 5/64 of them represent numbers. What fraction of the ASCII table remains to be used for other characters?   _____

16. A computer has 1,024 MB of RAM. If 1/8 of the RAM is being used for an on-line video, and 1/2 of the RAM is reserved for a virtual machine, how much RAM is left for the operating system?   _____

17. A 480-GB hard drive is partitioned into four sections. The first three partitions take up 1/4, 3/16, and 1/3 of the available space. How many gigabytes are left for the last partition?   _____

18. Converting from FAT16 to FAT32 will free up about 1/5 of the space on an old 2,000-MB hard drive. If the hard drive is currently full, how much space will be used after the conversion to FAT32?   _____

19. A programmer can work 37 1/2 hours in a week before collecting overtime. According to this employee's time card, how many weekend hours can she work without overtime?   _____

Day	Hours Worked
Monday	5½ hours
Tuesday	8¼ hours
Wednesday	5¾ hours
Thursday	5¼ hours
Friday	7¾ hours
Saturday	
Sunday	

20. You need to cut some network patch cables of the following lengths: 15 1/3 feet, 17 5/8 feet, 41 1/2 feet, and 16 3/4 feet. How much will be left over from your 100-foot roll?   _____

Unit 8 MULTIPLICATION OF COMMON FRACTIONS

BASIC PRINCIPLES

Multiplying common fractions is, in some ways, easier than addition and subtraction because it is not necessary to have a common denominator. Simply multiply the numerators together, and then multiply the denominators together. If necessary, the answer can be reduced to its lowest terms.

Example: $3/4 \times 2/5$

$$\frac{3}{4} \times \frac{2}{5} = \frac{3 \times 2}{4 \times 5} = \frac{6}{20} = \frac{3}{10}$$

In some cases, it may be possible to simplify a problem before you multiply by using *cross reduction*. For example, if the numerator of one fraction and the denominator of another can both be divided by the same number, a reduction can be carried out before the multiplication process.

Example: $7/12 \times 3/4$

$$\frac{7}{12} \times \frac{3}{4} = \frac{7}{12(\div 3)} \times \frac{3(\div 3)}{4} = \frac{7}{4} \times \frac{1}{4} = \frac{7}{16}$$

It is possible that a problem will involve mixed numbers as well as fractions. It is necessary then to change any mixed numbers into improper fractions before attempting to multiply the fractions. To convert a mixed number (3 1/2) to an improper fraction, multiply the whole number (3) by the denominator (2), and add the product (6) to the numerator (1).

Example: $3\ 1/2 \times 2/3$

$$3\frac{1}{2} \times \frac{2}{3} = \frac{(3 \times 2) + 1}{2} \times \frac{2}{3} = \frac{7}{2} \times \frac{2}{3} = \frac{14}{6} = \frac{7}{3} = 2\frac{1}{3}$$

REVIEW PROBLEMS

1. Installing a software patch takes 2/3 of an hour. How many hours will it take to install the patch on 37 computers? **Hint:** To make a whole number into a fraction, simply place it over 1 (37/1). _____

2. A computer store is having a 1/3-off sale. How much would you save on a $396 CPU? _____

3. A 32-bit class "C" IP address uses 3/4 of its bits to represent the network address. How many bits are used?   _____

4. The banner on a brochure has four sections. If each is 1 3/4-inch wide, how wide is the banner?   _____

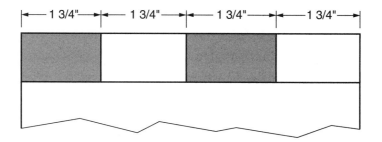

5. If half of computer users have a high-speed Internet connection, and 3/4 of them use it only to download music, what fraction is using their high-speed connection to download music?   _____

6. If an RFID tag is encased on both sides with a protective cover that is 2/7-mm thick, how thick is the cover encasing the RFID tag?   _____

7. A virus attempts to infect 3/4 of the computers in an office. If 2/3 of the computers are immune because they are running a current antivirus program, what fraction of the computers in the office will become infected?   _____

8. If area is found by multiplying length by width, what is the surface area on one side of a piece of 8 1/2-by-11-inch printer paper?   _____

9. Parity RAM uses 1/9 of its storage capacity for error checking. How much storage capacity would be used on 1/4 of a gigabyte of parity RAM?   _____

10. On a Windows 98 machine, the upper memory region takes up 3/8 of the first megabyte of RAM. If the motherboard BIOS occupies 1/3 of the upper memory region, what fraction of the first megabyte of RAM does it use?   _____

11. You can see 56 wireless networks from your laptop, and 3/8 of them are unsecured. How many unsecured networks do you see?   _____

12. Spam comprises 1/3 of all the messages received by your mail server. If a spam filter can delete 7/8 of the spam, what fraction of the remaining mail is spam?   _____

13. A standard 19-inch 1U rack-mountable server is 1 3/4-inch thick. How thick is a 5U backup power supply if it is five times as thick?

_____

14. If a motherboard is running at 133 1/3 MHz, and the CPU is running at a 12 1/2 multiple, how fast is the CPU running?

_____

15. Is an old bracket for 5 1/4-inch drives wide enough to hold two 3 1/2-inch hard drives side by side?

_____

16. A Web page is 640 pixels wide. If a frame on the left-hand side takes up 2/5 of the page, what is the frame's width?

_____

17. Heavy network traffic is making it necessary to re-send 1/3 of the packets. Out of those, another 1/3 must be re-sent a second time. What fraction of packets has to be re-sent twice?

_____

18. If 3/4 of all college students have a credit card, and 2/5 of all colleges accept credit card payment over the Internet, what fraction of college students could choose to pay tuition on-line?

_____

19. A programmer works 7 3/4 hours a day at $32 per hour. How much will he make in a five-day week?

_____

20. For data backup you have a minicartridge with the following measurements: 3 1/4 inches by 2 1/2 inches by 2 3/5 inches. Find the product to determine the volume of the minicartridge.

_____

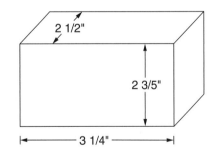

2 1/2"

2 3/5"

3 1/4"

Unit 9 DIVISION OF COMMON FRACTIONS

BASIC PRINCIPLES

Dividing common fractions is very similar to multiplying them, except that the divisor must first be inverted. After that, the two fractions are multiplied, and the product is reduced to its lowest terms.

Example: Divide 2/3 by 1/4.

$$\frac{2}{3} \div \frac{1}{4} = \frac{2}{3} \times \frac{4}{1} = \frac{8}{3} = 2\frac{2}{3}$$

REVIEW PROBLEMS

1. A CPU fan manufacturer recommends using 3/4 of a gram of heat-sink paste during installation. How many grams of heat-sink paste would be required to install 20 CPU fans? _____

2. A can of pressurized cleaning air lasts a technician 3 1/2 days. How many cans of air will she require in a 56-day period? _____

3. How many 3 1/2-foot patch cables can be cut from a 100-foot roll? _____

4. A software installation takes 1/4 hour. How many installations can be completed in a 7 1/2-hour day? _____

5. A bench technician is paid $750 for 37 1/2 hours of work. What does the bench technician earn per hour? _____

6. If three hard drives are using 132 5/8 watts, what is the number of watts used per hard drive? _____

7. A computer sales company finds that two out of five customers send in their warranty cards. If the company receives 1,890 warranty cards in the mail, how many computer systems were sold? _____

8. The director of an IT department finds she spends 3/8 of the budget on capital expenses. If the capital expenses for one year are $44,658, what is the total budget? _____

9. A hard drive rotates 15 times in 1/8 of a second. How many times does it rotate in a minute? **Note:** A second is 1/60 of a minute. _____

10. A computer is running 12 essential processes. If the essential processes represent 3/16 of the total, how many processes are running on the computer? _____

11. A new LCD monitor uses only 3/8 as much power as the CRT model it will replace. If the new monitor uses 48 watts of power, how much did the old one use? _____

12. A software analyst takes 4 1/4 days to debug a program. If debugging only takes 2/3 as long as programming, how long did the original programming take? _____

13. A CPU runs at a frequency of 3,600 MHz with a 4 1/2 times multiplier. What is the motherboard frequency? _____

14. A modem has a data transfer rate of 7 kilobytes per second. If a bit is 1/8 of a byte, what is the modem's data transfer rate in kilobits per second? _____

15. If an MP3 file is only 1/12 the size of the same file in an uncompressed format, what size file would have been used to create an 11 3/4-MB MP3 file? _____

16. A Web designer downloads a 113,400-KB file in 3/4 of an hour. What is the download rate in kilobytes per second? **Note:** A second is 1/60 of a minute, and a minute is 1/60 of an hour. _____

17. A 1 2/5-TB (terabyte) RAID array is divided into 3 equal partitions. How large is each partition? _____

18. A support technician re-images 64 computers in an 8 1/4-hour day. How many computers can she re-image in an hour? _____

19. For most personal computers, a bit is 1/4 of a nibble, a nibble is 1/2 of a byte, and a byte is 1/4 of a word. How many bits are in three words? _____

20. A USB 2.0 cable can be up to 16 1/2 feet long, which is 1 2/3 longer than USB 1.1 cable. What is the maximum length of a USB 1.1 cable? _____

Unit 10 COMBINED OPERATIONS WITH COMMON FRACTIONS

BASIC PRINCIPLES

Use what you have learned about addition, subtraction, multiplication, and division of common fractions to solve the following practical problems.

REVIEW PROBLEMS

1. Moore's Law says that storage capacity doubles every 1 1/2 years. If an entry level hard drive is 200 MB today, how large will one be in six years? _____

2. A Web designer gets time and a half for each hour she works over 40 hours in a week. If she normally makes $26 per hour, how much will she make in a 58-hour week? _____

3. A new CPU cooling fan promises to reduce the CPU temperature by at least one-third. Did the cooling fan perform as claimed if the CPU temperature dropped from 57 degrees to 39 degrees Celsius after the installation? _____

4. Windows represents used hard drive space with a pie chart. Suppose you have a computer with two hard drives. If one 120-GB hard drive looks 3/4 full, and the second 120-GB drive is only 1/3 full, what is the total amount of free space left on the computer? _____

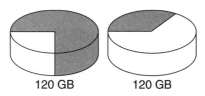

120 GB   120 GB

5. A 24-port network switch is used to connect the computers in three different offices. If 1/3 of the connections go to one office and 3/8 of the connections go to another, how many connections are left for the third office? _____

6. You have saved some pictures onto CDs. The CDs are filled according to the following chart. What is the minimum number of CDs required to back up all of your pictures?   _____

CD Label	Amount Used
2004	1/3
2005	1/2
2006	7/8
2007	4/5

7. You start scanning your hard drive for errors and then decide to go for lunch. You come back after 1/2 an hour, and the process is only 2/5 completed. How long will the whole scan take?   _____

8. Of all computer game players, 1/8 are men over age 50, and 2/5 are women. What fraction represents men 50 years old and under?   _____

9. A RAID-5 setup is striped across seven 200-GB hard drives to form a single volume. How large is the volume if it uses up 6/7 of the storage capacity?   _____

10. A computer has 1,024 MB of RAM. If 1/8 of the RAM is used for onboard video, and 1/2 of the RAM is reserved for a virtual machine, how much RAM is left?   _____

11. A computer has a 240-GB hard drive. If the extended partition makes up 1/3 of the drive and the second logical drive accounts for 1/4 of the extended partition, how large is the second logical drive?   _____

12. A computer has four slots for memory modules. If 1/2 of the slots contain 256-MB DIMMs, 1/4 of the slots have 512-MB DIMMs, and 1/4 of the slots are empty, how much RAM does the computer have?   _____

13. A KVM switch allows several servers to be connected to a single keyboard, video, and mouse. If 1/3 of 12 servers are connected to a 2-port KVM and the rest are connected to a 4-port KVM, how many KVM switches are being used?   _____

14. How much faster does a 7,200-rpm hard drive spin than a 5,400-rpm drive? Give the answer as a fraction.   _____

15. A project manager has 3 1/2 days to debug 10,000 lines of code. If one programmer can debug 1,000 lines per day, how many programmers must be assigned to the job?   _____

16. The potential speed of a computer is reduced by one-third because it is running 45 unnecessary processes. If 3/5 of those processes are removed, at what fraction of the potential speed will the computer now run? **Note:** Assume each process has an equal effect on the potential speed.   _____

17. A wireless router goes on sale at 1/3 off. If the sale price is $38, what was the original price?   _____

18. A computer has 2 1/2 GB of RAM. If 1/2 of the RAM is split among three virtual machines, how much RAM is reserved for each virtual machine?   _____

19. A computer spends 1/12 of its time switching between processes. If the time it takes to switch between processes could be reduced by one-third, what fraction of its time would be spent on switching?   _____

20. A small company decides to lease a fractional T1 line. If it leases 3/8 of the twenty-four 64-Kbps channels, how fast is the connection?   _____

Decimal Fractions

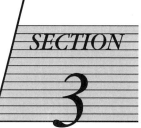

Unit 11 ADDITION OF DECIMAL FRACTIONS

BASIC PRINCIPLES

A decimal fraction is simply a common fraction whose denominator is expressed as in terms of tens or "a power of 10" such as 10, 100, or 1,000. Decimal fractions use a *decimal point* to separate a whole number from a fraction. The number to the left of the decimal point represents a whole number while the number to the right of the decimal point represents a fraction.

Another way to think about it is that the decimal point is the center, where any numbers to the left of the center are whole tens, hundreds, thousands, and so forth, and any numbers to the right of the center are fractions (less than 1) and are expressed as tenths, hundredths, thousandths, and so forth. The whole and fractional digits are joined by the decimal point to form a decimal fraction.

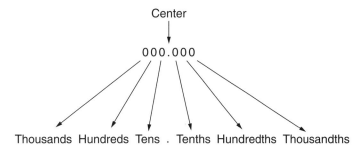

Center

000.000

Thousands Hundreds Tens . Tenths Hundredths Thousandths

Example: Express the mixed number 5 4/10 as a decimal fraction.

5 + 4/10 = 5.4

The number to the left of the decimal point (5) is the whole number and is written down first. The numerator (4) goes in the "tenths" column just to the right of the decimal point.

Example: Express the number 0.372 as a fraction.

0.372 = 372/1,000

In this case, the number on the right of the decimal point contains three digits, so it is expressed as a fraction of 1,000. The denominator is multiplied by 10 for each digit (10 × 10 × 10 = 1,000).

Adding decimal fractions is often easier than adding common fractions because a common denominator is found by simply aligning all of the decimal points vertically.

Example: Find the sum of 9.7 + 2

$$
\begin{array}{r}
9.7 \\
+\ 2. \\
\hline
11.7
\end{array}
$$

Example: Find the sum of 21.2751 + 5.4 + 0.372

$$
\begin{array}{r}
21.2751 \\
5.4000 \\
+\ 0.3720 \\
\hline
27.0471
\end{array}
$$

REVIEW PROBLEMS

1. A hard drive is divided into three partitions. One is 3.4 GB, another is 26.5 GB, and the third is 15.7 GB. How large is the hard drive?   _____

2. You need to store four different media files. They are 121.6 MB, 35.8 MB, 116.0 MB, and 25.4 MB in size. How much room is required to store all four files?   _____

3. Upgrading a computer to run a new operating system requires a new video card for $89.95 and some additional RAM that costs $115.57. What is the cost of the upgrade?   _____

4. A service technician charges $114.76 for parts and $76.65 for labor to upgrade a computer. What is the total cost of the upgrade?   _____

5. A Mini-ATX motherboard measures 11.2 inches by 8.2 inches. Find the sum of all four sides to determine the perimeter? _____

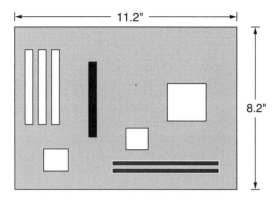

6. A wireless connection is added to a computer with the following parts. What is the total cost? _____

Item	Cost
Wireless Router	$49.99
6-foot Cat 5 Cable	$5.47
9-foot Cat 5 Cable	$7.55
Wireless Adapter Card	$18.75

7. Your computer is running three processes. They use 34.7 MB, 117.0 MB, and 64.1 MB of RAM. What is the total amount of RAM used by these processes? _____

8. Several SCSI devices are daisy-chained together with three cables. If the cables measure 3.1 m, 0.8 m, and 2.5 m, what is the total length of all three cables? _____

9. The manager of a software development team lays out the following
    schedule. How long will the project take?   _____

Phase	Time Required
System Concept	2.0 days
System Design	3.5 days
Detailed Design	4.2 days
Coding	5.7 days
Testing	7.5 days

10. A user downloads query results from an SQL server that are 12.34 MB,
    15.22 MB, 54.80 MB, and 11.0 MB in size. What is the total size of the
    downloaded files?   _____

11. Two LAN cables that measure 15.8 m and 56.7 m are connected by a
    repeater. What is the total length of the cable run?   _____

12. One box of hard drives weighs 16.75 lb (pounds) and another weighs
    26.66 lb. What would both boxes weigh if lifted together?   _____

13. A DVD-5 holds 4.7 GB of data, and a DVD-9 holds 8.5 GB. How much
    data could you save using one of each?   _____

14. The components of a server and their weights are listed in the table
    below. What is the total weight of the server?   _____

Component	Weight
Chassis	67.8 lb
System Board	1.2 lb
Hard Drives	7.8 lb
Tape Backup	2.1 lb

15. A second hard drive requires 37.5 watts of power, and an extra DIMM of RAM needs 40.2 watts. What will be the total power requirement of this upgrade?   _____

16. A laptop comes with a lithium ion battery that lasts for 2.7 hours, and a DMFC (direct methanol fuel cell) that lasts 5.4 hours. How long could you run your laptop off of these two batteries?   _____

17. Three cables measuring 35.7 cm, 18.1 cm, and 47.9 cm are connected. Find the total length.   _____

18. The traceroute command reveals the time spent at three different hops to be 10.67 ms, 6.08 ms, and 4.66 ms. What is the total time?   _____

19. Find the total size of the following files: 15.67 MB, 296.00 MB, 18.88 MB, and 10.50 MB.   _____

20. A support technician logs the following miles in a week. What is the total distance driven?   _____

Day	Distance
Monday	36.21 mi
Tuesday	12.00 mi
Wednesday	74.09 mi
Thursday	66.20 mi
Friday	79.98 mi
Saturday	0 mi
Sunday	0 mi

Unit 12 SUBTRACTION OF DECIMAL FRACTIONS

BASIC PRINCIPLES

Subtraction of decimal fractions follows the same rules that apply to addition. When subtracting, however, place the second number under the first one and align the decimal points.

Example: 72.382 – 16.663

$$\begin{array}{r} 72.382 \\ -16.663 \\ \hline 55.719 \end{array}$$

Example: 14 – 9.6

$$\begin{array}{r} 14. \\ -\ 9.6 \\ \hline 4.4 \end{array}$$

REVIEW PROBLEMS

1. A dual voltage processor has an I/O voltage of 3.4 V and a core voltage of 2.8 V. What is the difference between the two? _____

2. A hard drive is 155.28 GB in size. If it has three partitions of 24.11 GB, 57.81 GB, and 47.05 GB, how much unpartitioned space is left on the drive? _____

3. The capacity of a volume is 9.68 GB. If 3.15 GB of space is left free, how much of the volume is filled with data? _____

4. An overheating CPU is running at 76.5 degrees Celsius. A larger heat sink and fan would drop the temperature by 20.8 degrees. Would this bring the temperature below 55.4 degrees? _____

5. The acceptable range of the +12-V source from a PC power supply is anywhere from +10.8 V to +13.2 V. How large is this range? _____

6. A CPU is currently running at 76.8 degrees Celsius. Adding a larger heat sink will reduce the temperature by 12.8 degrees, and adding a case fan will drop the temperature another 5.6 degrees. What will be the resulting CPU temperature?   _____

7. In the diagram below, is there a space of at least three feet between the server racks?   _____

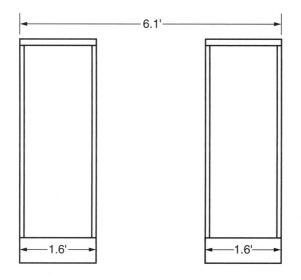

8. How much longer is a 5.256-inch 240-pin DIMM than a 4.255-inch 72-pin SIMM?   _____

9. A CMOS battery is rated at 3.6 V but is only measured at 2.9 V. How much below the rating is the actual measured output?   _____

10. An uncompressed file is 15.54 MB in size. When compressed, it is only 9.82 MB. How much room is saved with the compression?   _____

11. A network technician has a piece of Cat 5e cable that measures 43.7 m in length. If he cuts off patch cables that are 1.1 m, 2.7 m, 4.1 m, and 1.9 m, how much will be left?   _____

12. Your backup will take 45.11 MB of space but will overwrite an old backup taking 37.45 MB. How much new space will be used?   _____

13. You get a raise from $16.85 per hour to $21.25 per hour. How much more will you be making for each hour you work?   _____

14. The cost of a stick of RAM drops from $144.69 to $120.38. How much will you be saving?   _____

15. An empty flash drive holds 512 MB of data. How much is left after you save files that are 15.27 MB, 18.90 MB, and 6.23 MB?   _____

16. A graphic artist downloaded 116.2 MB of images from his digital camera but then deleted 87.9 MB of them. How many megabytes of files did he keep?   _____

17. The PCIe bus transfers data at 4.0 GB/s, while the 8× AGP bus works at 2.1 GB/s. How much faster is the PCIe bus?   _____

18. A portable flash memory device contains 984.57 MB of files, but you remove files that are 15.82 MB, 1.12 MB, 0.88 MB, 17.00 MB, and 31.11 MB. How many megabytes of files remain on the drive?   _____

19. It takes your computer 15.8 seconds to load the BIOS during a cold boot. If a warm boot saves 3.4 seconds by skipping parts of the POST, how long will it take to load the BIOS during a warm boot?   _____

20. You want to backup the folders listed below. Will all of these folders fit onto one CD that holds 800 MB?   _____

Folder	Size
Accounts	257.87 MB
Sales	111.15 MB
Purchases	331.03 MB
Expenses	99.84 MB

Unit 13 MULTIPLICATION OF DECIMAL FRACTIONS

BASIC PRINCIPLES

Multiplication of decimal fractions involves the same procedure used for multiplying whole numbers—except that the placement of the decimal point in the product must be determined. The number of decimal places in the product (counted from right to left) matches the total number of decimal places in all the decimal fractions being multiplied.

Example: 2.67×1.5

$$
\begin{array}{r}
2.67 \\
\times\ \ 1.5 \\
\hline
1335 \\
2670 \\
\hline
4.005
\end{array}
$$

Because the first number (2.67) has two decimal places, and the second number (1.5) has one decimal place, the total number of decimal places is three. The decimal is then added three spaces from the right-hand side of the product.

Example: 0.534×2

$$
\begin{array}{r}
0.534 \\
\times\ \ \ \ 2 \\
\hline
1068 \\
\hline
1.068
\end{array}
$$

In this case, the first number (0.534) has three decimal places while the second number (2) has none. The total number of decimal places is three. The decimal place is therefore added three spaces from the right-hand side of the product.

REVIEW PROBLEMS

1. A motherboard has an 800-MHz frontside bus and runs at a 4.5× multiplier. Find the product to determine the clock speed of the processor.   _____

2. The airflow needed to cool a motherboard depends on the surface area of the motherboard. What is the surface area on one side of a 12-by-9.6-inch ATX motherboard?   _____

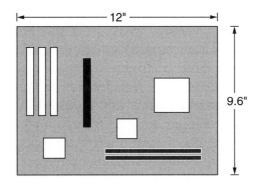

3. A single-sided, single-layer DVD can hold 4.7 GB of data. How much data can be stored on a double-sided, single-layer DVD?   _____

4. A DVD is 1.2 mm thick. How thick is a spindle of 50 DVDs?   _____

5. A monitor requires 1.8 A at 120 V. How many watts does it require? **Note:** W (watts) = V (volts) × A (amperes).   _____

6. A Pentium IV processor has a system bus speed of 800 MHz and a 4.75× multiplier. What is the processor speed?   _____

7. The original data transfer rate of a CD-ROM is 0.15 MBps. What is the rate of a new 56× CD-ROM?   _____

8. A Windows administrator creates a striped volume across two 237.64 GB disks. How large is the volume?   _____

9. A programmer is paid $36.50 per hour. How much does she make in a 7.5-hour day?   _____

10. A small office purchases 23 software licenses for $27.34 each. How much is the total purchase of licenses?   _____

11. You get a raise of $1.15 per hour. How much extra will you earn in a 38-hour work week? _____

12. If 0.3048 m is equal to 1 foot, is a 620-foot network cable less than 200 m? _____

13. A flat-screen monitor measures 16.5 inches by 12.2 inches. What is the surface area of the monitor? _____

14. A tech support assistant downloads a driver file at 0.58 MBps. If the download takes 38.5 seconds, how large was the driver? _____

15. An organization has a 35.4 MB database stored on the server, and users must run the database by downloading it to their local machines. If 14 users need to use the database, what is the total size of the files to be downloaded? _____

16. A case fan moves 68.23 cfm (cubic feet per minute). How much air will it move in five minutes? _____

17. A UPS is rated for 750 VA (volt-amps) and has a power factor of 0.6 watts per VA. Calculate how many watts the UPS supplies by multiplying the VA rating by the power factor. _____

18. The maximum distance for STP (shielded twisted pair) cable is 328.08 feet or 100 meters. How many meters are in one foot? _____

19. If one backup tape can store 3.2 GB of data, how much data can be stored on seven tapes? _____

20. A programmer is writing a function to calculate the circumference of a circle using the equation Circumference $= 2 \times \pi \times$ radius and wants to test the results. What result should he expect with a radius of 5.47 and a value for π of 3.14? _____

Unit 14 DIVISION OF DECIMAL FRACTIONS

BASIC PRINCIPLES

Division of decimal fractions begins by placing the divisor to the left of the dividend. The divisor must then be turned into a whole number by moving the decimal place all the way to the right. The decimal point is also moved an equal number of places in the dividend to create an equivalent fraction. The decimal point in the quotient will be aligned with the decimal point on the dividend.

Example: $6.2 \div 0.4$

The divisor (0.4) was turned from the decimal number into a whole number (4) by moving the decimal one place to the right. In order to keep the dividend equivalent in value to the new divisor, the decimal point was also moved one place to the right (6.2 became 62).

REVIEW PROBLEMS

1. A liquid cooling unit for a CPU requires 355 ml of fluid. If one container of cooling fluid holds 52.7 ml, how many containers are required?   _____

2. A type-I PC card is 3.3 mm thick, and a type-III PC card is 10.5 mm thick. Could three type-I PC cards fit into a case designed for a single type-III card?   _____

3. All things being equal, how many times faster is a 3.4 GHz processor than a 1.8 GHz processor?   _____

4. If a double-sided DVD can hold 9.4 GB of data, how much data can fit onto a single-sided DVD?   _____

5. A spool of network cable is 200 feet long. How many 4.75-foot cable lengths can be cut from the spool?   _____

6. A private software developer copies 10 CDs in 1.25 hours. How many CDs can be copied in a 7.5-hour day?   _____

7. A network technician buys a box of 500 RJ-45 connectors for $30.00. How much does each connector cost? _____

8. You replace the toner on your laser printer after 1,568 pages. If the toner cartridge was purchased for $188.16, what was your cost of printing per page? _____

9. A 3U rack-mountable power supply is 5.25 inches thick and is three times thicker than a 1U device. How thick is a 1U device? _____

10. A CPU is running at 2.8 GHz with a 7× multiplier. What is the motherboard speed? _____

11. If the 16× PCIe bus has a transfer rate of 4 GBps, what is the transfer rate of the 1× PCIe bus? _____

12. A 100-foot spool of Cat 5 cable costs $15.95 and weighs 3.5 pounds. Find the cost per pound. _____

13. A computer has a 568.68-GB volume spread over three equal hard drives. What is the size of each hard drive? _____

14. A hard drive transfers 3.990 GB in one minute. How much data does it transfer per second? _____

15. A laser printer can print 6.2 pages per minute. How long will it take the printer to print 93 pages? _____

16. A 3.8-GB file is too large to fit on a single 0.7-GB CD so it must be spanned across several CDs. How many CDs will be required? _____

17. A Web master uploads 61.70 MB of new content to her Web site in one week. If an equal amount is uploaded each day, how much data is uploaded per day in a five-day week? _____

18. Your old video card could render 32.16 fps (frames per second) in a particular benchmark program. How many times faster is your new video card if it can display 80.40 fps? _____

19. If a 340 GB hard drive costs $400.00, what is cost per gigabyte? _____

20. You are copying old files from 1.44 MB floppy disks. An 80,000 MB external hard drive will hold the data from how many disks? _____

Unit 15 COMBINED OPERATIONS WITH DECIMAL FRACTIONS

BASIC PRINCIPLES

Now use what you have learned about addition, subtraction, multiplication, and division of decimal fractions to solve the following practical problems.

REVIEW PROBLEMS

1. The original 16-bit ISA bus transferred data 2 bytes per cycle, ran at 8.33 MHz (million cycles per second), and could only send data every second cycle. At what rate did it transfer data in MB/s? _____

2. A video card draws 0.5 A at 5.0 V, 2.0 A at 3.3 V, and 0.1 A at 12 V. Find the sum of all three products to calculate the watts used by the video card. **Note:** Watts is calculated by multiplying volts by amps. _____

3. A Firewire port has a maximum speed of 1.2 Gbps (gigabits per second). How many GB (gigabytes) could it transfer in one minute? **Remember:** 8 bits = 1 byte. _____

4. An OC-3 fiber backbone runs at 0.1536 Mbps and is three times faster than an OC-1 connection. How fast would you expect an OC-24 connection to be? _____

5. A network administrator has $12,000 to upgrade all of the computers in an office with 40 identical systems. If new DVD burners cost $57.89 and larger hard drives cost $134.52, how much can she spend, per computer, on more RAM? _____

6. A case fan exchanges 0.2 cubic feet of air per second. How many seconds will it take to exchange all the air in a case that measures 0.58 feet by 1.32 feet by 1.40 feet? _____

7. The known failure rate for a new CPU is 0.75 per 1,000 units. If you purchased 12,000 CPUs, how many would you expect to have to return? _____

8. A 7.2-megapixel camera will take a photograph that can be enlarged to 8 by 10 inches without pixilation. How many megapixels would you need if you only wanted a 3- by 4-inch photo? _____

9. If a computer power supply is providing 9.2 A at 12 V, 46.1 A at 5 V, and 52.7 A at 3.3 V, how many watts is it supplying? **Remember:** W (watts) = V (volts) × A (amps). _____

10. An adware removal program can scan an average of 100 files in 0.28 seconds. How long will it take to scan 12,688 files? _____

11. A computer assembler is paid $15.50 per hour but gets time and a half on the weekends. How much will she earn for the work on this time sheet? _____

Day	Hours Worked
Monday	5.6 hours
Tuesday	3.4 hours
Wednesday	7.1 hours
Thursday	8.0 hours
Friday	0
Saturday	7.7 hours
Sunday	4.0 hours

12. A network technician runs four cables of equal length from the server room to an office. If there are only 15.6 feet of cable remaining from the original 100-foot piece, how long is each of the newly installed cables? _____

13. A computer animator is running a program to render a transparent object in a 10.4-MB video file. After 10 minutes, the program has only completed 0.8 MB of the file. How long will the program take to complete? _____

14. A server's hard drive has an access time (the time it takes to start loading a file) of 4.5 milliseconds (ms), and a data transfer rate (the time it takes to read a file) of 0.8 MB per millisecond. How long will it take the server to read two files that are 10 MB and 15 MB in size? _____

15. The power requirements for a CD-ROM are listed as 1.3 A at 5 V and 1.5 A at 12 V. Find the sum of these products to determine the total number of watts required. **Remember:** (W = V × A). _____

16. If a DSL Internet connection takes 11.26 seconds to download a 2-MB file, how long would it take to download a 0.87-MB file?   _____

17. Server components are 1.75 inches high (for each U). What would be the height of a 1U server, a 3U tape backup unit, and a 5U UPS, if they were stacked on the same rack?   _____

18. A bench technician gets paid $19.85 per hour and works 75 hours in every two-week pay period. If the deductions from her paycheck total $375.68, what is her net pay?   _____

19. If you pay $39.00 for three spindles of 50 CD-R disks, how much did each disk cost?   _____

20. The cost of hard drives of equal size drops by half every 1.5 years. If a 200-GB hard drive currently costs $105.80, what would you expect to pay for a 200-MB drive three years from now?   _____

Statistics and Estimates

 ## Unit 16 *PERCENT AND PERCENTAGES*

BASIC PRINCIPLES

Percent is another kind of fraction where the denominator is hidden and has a value of 100. The word "percent" and the symbol % literally mean "per one hundred." If a network technician has 100 RJ-45 connectors and uses 25 of them, you can say that 25/100 or 25% have been used.

Example: Express 0.55 as a percent.

To convert a decimal fraction to a percent, move the decimal point two places to the right and add a percent sign (%) after the number.

$$0.55 = 55\%$$

Example: Express 2.6% as a decimal fraction.

To convert a percent to a decimal fraction, move the decimal point two places to the left and remove the percent sign.

$$2.6\% = 0.026$$

Example: Express the common fraction 3/4 as a percent.

To convert a common fraction to a percent, divide the numerator by the denominator and then convert the resulting decimal fraction to a percent.

$$\frac{3}{4} = 3 \div 4 = 0.75 = 75\%$$

Example: Express 62.5% as a common fraction.

To change a percent into a common fraction, change the number into a decimal fraction and then reduce the fraction to the lowest terms.

$$62.5\% = 0.625 = \frac{625}{1000} = \frac{625(\div 125)}{1000(\div 125)} = \frac{5}{8}$$

Example: The number 9 is what percent of 60?

To calculate percent, divide the part (9) into the whole (60) to first get a decimal fraction, then convert the decimal fraction into a percent.

$$\frac{9}{60} = 9 \div 60 = 0.15 = 15\%$$

REVIEW PROBLEMS

1. A junior programmer receives $975.00 a week and gets a 4% raise. How much more does she get after the raise?   _____

2. If 2.5% of the hard drives you received have bad sectors, how many have to be returned out of a batch of 320?   _____

3. A programmer makes 2 mistakes on 45 lines of code. Find the percentage of mistakes to the nearest tenth of a percent.   _____

4. A computer repair company charges a 25% markup on all parts. What is the consumer price (to the nearest cent) on a stick of RAM that costs the company $67.43?   _____

5. A laptop battery lasted for 150 minutes when it was new, but now it only lasts 120 minutes. Find the percentage decrease in battery life.   _____

6. Of all Internet users, 25% use the Internet for downloading music. Convert this number to a common fraction.   _____

7. Three tenths of teenagers use the Internet to search for a job. Express this number as a percent.   _____

8. If 313 million users represent the 31% who speak English, what is the total number of Internet users?   _____

9. Your computer's power supply can provide 450 watts. If it is working at 80% capacity, how many watts are being drawn?   _____

10. The internal case temperature on a computer normally runs at 85 degrees. If a case fan drops this temperature by 12%, what is the new running temperature to the nearest degree?   _____

11. If 44% of e-mail originates in the United States what percentage of e-mail originates outside of the United States? _____

12. A computer technician charges $375.00 to do a computer upgrade. The cost of materials accounts for 65% of the bill. How much was charged for labor? _____

13. A 37.8-GB hard drive is 12.1 GB full. What percentage of the hard drive is full (to the nearest tenth of a percent)? _____

14. One program is responsible for 8% of the 50 processes running on a computer. How many processes is this program running? _____

15. A network administrator is responsible for 114 computers, but only 87 are working. What percent are working? _____

16. A blank CD-RW will hold 650 MB of data. What percent of the space is left after you store a 520-MB backup file? _____

17. A company spends $12,000 out of its $85,000 capital budget on purchasing new computers. To the nearest tenth of a percent, what percent of the capital budget has been spent? _____

18. An antivirus program is 98.5% effective in preventing viral infections on a computer. If 17,366 people use this program on their home computers, how many computers might become infected? _____

19. Three quarters of 1% of RJ-45 connectors are found to be defective. Express this number as a decimal fraction. _____

20. You are taking an on-line accreditation test, and you don't have time to answer 6 of the 75 questions. What is the maximum you can score on the test to the nearest percentage point? _____

 *Unit 17* **INTEREST AND DISCOUNTS**

BASIC PRINCIPLES

Two ways that percentages are often used are interest and discounts. When money is borrowed, the initial amount borrowed is known as the *principal,* and the fee for borrowing this money is called the *simple interest*. This simple interest is usually calculated on a yearly basis and is expressed as a percentage of the principal. Most loans are made for a specific time period or term. When the term of the loan is expired, the principal must be repaid as well as the interest.

In many cases, a lender might charge interest not only on the outstanding principal but also on the outstanding interest. This is called *compound interest*. The following examples and problems deal only with simple interest.

Example: You borrow $8,000 at a rate of 7% simple interest for one year. How much interest must be paid after 1 year?

Convert the interest rate (7%) to a decimal fraction (0.07) and then multiply this number by the principal ($8,000) and the term (1 year) to determine the interest charge.

Interest = 0.07 per year × $8,000 × 1 year = $560

Example: Calculate the total to be repaid on a $1,000 loan with a 15% interest rate for a term of 2.5 years.

Interest = 0.15 per year × $1,000 × 2.5 years = $375

The interest is then added to the principal to determine the total amount of money to be repaid.

Total Amount = $1,000 + $375 = $1,375

The other type of percentage that we often encounter is a discount. Sometimes prices are marked down with a discount. A *discount* is a percent that is subtracted from the list price of an item or service.

List Price – Discount = Net Price

Example: A computer hard drive normally sells for $150 but is marked down with a 20% discount. What is the sale (net) price?

Begin by converting the 20% discount into a decimal fraction.

20% = 0.2

Then multiply the list price ($150) by the decimal fraction discount (0.2).

$$\$150 \times 0.2 = \$30$$

Finally, subtract the value of the discount from the original list price.

$$\$150 - \$30 = \$120$$

REVIEW PROBLEMS

1. The owner of a computer repair shop borrows $5,000 at 8% interest per annum and pays back the loan after one year. What is the interest charged?   _____

2. A network technician borrows $3,500 at 7.5% interest per annum to take a cabling course and doesn't pay back the loan for 3 years and 2 months. How much interest is charged?   _____

3. You borrow $250,000 at an interest rate of 14% to start up a software development company. What is it costing you each month to carry this debt?   _____

4. A Web designer travels to a conference and pays for the $700 flight with his credit card. How much interest will he pay if the credit card company charges 18.5% per annum and the loan is not repaid for 6 months?   _____

5. A programmer gets a $2,000 advance on her pay at 14% interest per annum and pays it back 4 months later. What is the total amount she has paid back?   _____

6. A call center manager borrows $50,000 to upgrade some computers. The money is borrowed at 12.5% interest per annum and not paid back for 18 months. What is the total cost of this loan?   _____

7. A computer retailer assembles computers with parts that cost $1,586 per computer. If the money to purchase these parts is borrowed at 16.5% interest per annum and it takes 3 months to sell the assembled computers, what is the total cost of each computer?   _____

8. An Internet service provider charges 29.5% interest per annum on outstanding bills. What would it cost to pay off a $750 Internet charge after four months?   _____

9. A company borrows $12,000 on which $1,680 interest is paid annually. What is the rate of interest?   _____

10. You borrow $10,000 to set up a Web hosting company, and you pay back a total of $13,000 after 24 months. What was the interest rate?   _____

11. A video card is marked down from $149.99 with a 20% discount. What is the sale price?   _____

12. A Web designer charges $45 an hour but offers a 10% discount for repeat customers. What hourly rate does she charge repeat customers?   _____

13. A programmer buys an ergonomic keyboard that usually sells for $69.95. What is the cost if she gets a 25% discount?   _____

14. What is the cost of an uninterruptible power supply if it has a list price of $264 with successive discounts (where additional discounts are off of the new discounted price) of 26% and 4%?   _____

15. A CAD program is on sale at 25% off, and there is an additional 3% discount if the buyer pays cash. If the program usually retails for $250, what would your net cost be if you bought it on sale with cash?   _____

16. A business purchases computers through a wholesale outlet at 65% of the retail cost. What percent of the retail price is saved?   _____

17. Which is a better deal: three successive 10% discounts or one 30% discount off the original price?   _____

18. A computer technician purchases $1,700 worth of equipment with successive discounts of 10% and 15%. What is the net cost?   _____

19. A company purchases 250 software licenses priced originally at $27.50 each, but the company negotiates the following discount schedule. What is the total cost of the software licenses?   _____

Quantity	Discount
First 1–10	None
Next 11–100	5%
Next 101–500	10%

20. Which is a better value: (1) ordering a package of 250 RJ-45 connectors for $95 at no discount or (2) ordering a package of 300 RJ-45 connectors for $115 at a 20% discount?   _____

Unit 18 *AVERAGES AND ESTIMATES*

BASIC PRINCIPLES

An *average* is a single value that is used to represent a group of values. A common form of average—called the *mean*—can be calculated by adding all of the values in a group and then dividing this sum by the number of values in the group.

Example: Three hard drives measure 120 GB, 80 GB, and 100 GB. What is the average size of these hard drives?

120 GB + 80 GB + 100 GB = 300 GB

$$\frac{300 \text{ GB}}{3} = 100 \text{ GB}$$

Often averages are used as a way of estimating cost. Estimates are used for approximating; it is understood that they are not exact.

Example: Estimate what it will cost to purchase 50 new computers if an average new computer costs $2,000.

50 computers × $2,000 = $100,000

The accuracy of an estimate can be calculated by taking the difference between the estimate and the actual amount and then dividing it into the original estimate, with the results being expressed as a percentage.

Example: It was estimated that a job would cost $1,200, but it ended up costing $1,400. What was the accuracy of the estimate?

$1,400 − $1,200 = $200

$$\frac{\$200}{\$1,200} = 0.17 = 17\%$$

REVIEW PROBLEMS

1. A wireless connection is measured at 27 Kbps, 14 Kbps, and 36 Kbps on three separate occasions. What is the average connection speed (rounding to the nearest Kbps)? _____

2. A programmer writes 560, 1,520, 764, and 833 lines of code on four consecutive days. What is the average number of lines of code produced per day?

   _____

3. A computer technician earns $35.50 and hour and works the hours listed below. What is the average daily earning for this technician?

   _____

Day	Hours
Monday	3 hours
Tuesday	5.5 hours
Wednesday	8.5 hours
Thursday	7 hours
Friday	8 hours

4. The network patch cables in a lab vary in length. Fifteen of them are 10-feet long, 10 of them are 15-feet long, and 5 of them are 30-feet long. What is the average length of a patch cable?

   _____

5. A Web design company creates 83 Web sites in its first year. What is the average number of Web sites created per month (rounded to a tenth)?

   _____

6. A laptop battery has lasted 2.3 hours, 1.8 hours, and 2.1 hours on its last three uses. Find the average battery life to the nearest tenth of an hour.

   _____

7. The CPU temperature is measured on three separate occasions at 35°C, 41°C, and 28°C. What is the average temperature?

   _____

8. CPU utilization is measured at 37%, 65%, 70%, 33%, and 49%. Does this exceed an average of 60% utilization?

   _____

9. A programmer earns $1,400.00; $1,312.50; $1,400.00; and $945.00 weekly over a four-week period. If the rate of pay is $35 per hour, find the average number of hours worked per week.

   _____

10. Several IT students are taking an on-line accreditation exam. If two students score 100%, one student scores 95%, three students get an 80%, and one student gets a 75%, what is their average score?

   _____

11. A help desk technician works 8 hours on Monday, 6 hours on Tuesday and Thursday, and 6.5 hours on Wednesday. How many hours must be worked on Friday to average 8 hours per day?   _____

12. If a 200-GB hard drive costs $115 and a 300-GB costs $172.50, what is the average cost of a hard drive per GB?   _____

13. A technician can assemble a computer in an average of 18 minutes. Estimate how many hours it will take two technicians to assemble 100 computers.   _____

14. It takes an average of $420 to upgrade a computer to run a new operating system. Estimate (to the nearest $1,000) what it would cost to upgrade an office with 120 computers.   _____

15. An office has 75 computers, each with an average of 750 MB of RAM. If RAM costs $0.70 per MB, estimate what it would cost to upgrade the whole office to 1,024 MB of RAM.   _____

16. A network technician estimates that two 1,000-foot spools of Cat 5e cable should be enough to wire a building with 8 offices. If the cable costs $123.50 per spool what is the estimated cost per office?   _____

17. A senior programmer estimates that it will take three 8-hour days to debug 2,000 lines of code. How many lines of code does the programmer think she can debug per hour?   _____

18. A network wiring job ended up costing $2,280 for 40 connections. A technologist originally estimated that it would cost about $60 per connection. How much over or under was the estimate?   _____

19. A call center manager estimates that $29,000 will have to be spent to upgrade the phone system. The cost ends up being $32,122. Find the percent of accuracy of the estimate.   _____

20. A programmer estimates that debugging will take eight days, but the job is completed after only six. Find the percent of accuracy of the estimate.   _____

Units and Notation

Unit 19 EXPONENTS

BASIC PRINCIPLES

Exponents are a shorthand way of describing the process of repeated multiplication. Suppose you wanted to multiply 3 by itself 4 times. You write this as $(3 \times 3 \times 3 \times 3)$ or you could write 3^4 where the small number 4 indicates the number of times to multiply 3 by itself. The small number is called the exponent or *power*.

Example: Raise the number 7 to the third power.

$$7^3 = (7 \times 7 \times 7) = 343$$

Numbers that are powers of 2 are quite common when dealing with computers because computers perform operations in binary (base 2). For example, the number of unique numbers that can be stored in memory is calculated by taking 2 "to the power of" the number of bits being used.

Example: How many unique numbers can be stored using 8 bits?

Because $2^8 = 256$, we know that 256 unique numbers can be stored. They range from 0 to 255.

REVIEW PROBLEMS

1. Powers of 2 often occur when measuring computer components such as memory. Find the following powers of 2.

   a. 2^1 _____   f. 2^6 _____

   b. 2^2 _____   g. 2^7 _____

   c. 2^3 _____   h. 2^8 _____

   d. 2^4 _____   i. 2^9 _____

   e. 2^5 _____   j. 2^{10} _____

2. A kilobyte (KB) is equal to 2^{10} bytes. Express this number as a whole number.   _____

3. A 16-bit sound card captures 2^{16} levels of sound. Express this number as a whole number.   _____

4. How many colors are available in 24-bit color scheme?   _____

5. A megabyte (MB) is equal to 2^{20} bytes. Express this as a whole number.   _____

6. If the size of an average hard drive doubles every two years, how many times larger will hard drives be in ten years?   _____

7. A computer has a case fan whose radius is 5 cm. What is the area of the opening that houses the fan? Remember to also take the units (cm) to the power of 2 (square cm or cm^2). (**Assume:** Area = radius^2 × 3.14)   _____

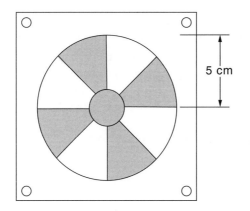

5 cm

8. A Micro-ATX motherboard measures 9.6 inches squared. What is the surface area of one side?   _____

9. If a CD label has a radius of 60 mm, what is its surface area? (Assume the label does not have a hole in the middle.)   _____

10. Protected mode uses 12 more address lines than the 20 used in real mode. If real mode has access to 1 MB of RAM, how much RAM can be addressed in protected mode? **Hint:** Protected mode used $12 + 20 = 32$ address lines. _____

11. Before the ATA/100 standard, 28 bits were used to address all of the half-KB sectors on a hard drive. What is the maximum size of one of these hard drives? _____

12. Floppy drives use a 12-bit file system to store data. Find how many unique numbers can be stored with 12 bits to determine how many separate allocation units are on a floppy drive? _____

13. A class "B" network is subnetted so that each network can accommodate 2^{11} hosts. Express this as a whole number. _____

14. How many new networks would be created if 6 bits (2^6) were "borrowed" from an IP address to make new networks? _____

15. A compact computer case measures 5 1/4 inches cubed. What is its volume? _____

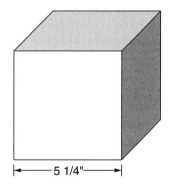

|←——— 5 1/4" ———→|

16. ASCII uses 7 bits (2^7) to address different characters. How many characters are in the ASCII table? _____

17. Memory locations are often written in the hexadecimal numbering system (counting with 16 symbols). How many memory addresses can be described with a 4-digit hex number? _____

18. There are 256^4 potential IP addresses. Express this as a whole number. _____

19. People can discriminate about 250^3 different colors. Express this as a whole number. _____

20. FAT16 breaks up a hard drive into 16^4 allocation units. Express this as a whole number. _____

Unit 20 *SCIENTIFIC AND ENGINEERING NOTATION*

BASIC PRINCIPLES

Exponents are a simple way of expressing numbers that are very small or very large. Suppose you want to express the number 150,000,000,000 without using so many zeros. This number is the same as $1.5 \times 100,000,000,000$—and 100,000,000,000 can more easily be expressed as the eleventh power of 10 (10^{11}). Expressing a number in this form is sometimes called *scientific notation*.

$$150,000,000,000 = 1.5 \times 10^{11}$$

Notice that every time you multiply a number by ten, the decimal point moves one space to the right, and every time you divide a number by ten the decimal point moves one space to the left.

Example: Express 64,000 in scientific notation.

$$64,000 = 6.4 \times 10^4$$

The decimal point moves four places to the left to become 6.4, which must then be multiplied by 10 four times in order to remain equivalent.

Example: Express the number 0.0000456 in scientific notation.

$$0.0000456 = 4.56 \times 10^{-5}$$

When the decimal point moves to the right, the exponent becomes negative. In this case the exponent of −5 indicates that 4.56 must be divided by 10 five times to remain equivalent.

Engineering notation is a specialized form of scientific notation where the powers are limited to multiples of 3 such as 10^3, 10^6, or 10^9. This allows engineers to deal with numbers that are very large (or very small) in a standardized way. These powers are often given a special name (such as thousand, million, or billion) or a special prefix (like kilo-, mega- or giga-). For example, instead of saying that a cable is 2,300 meters long, they would write the length as 2.3×10^3 meters or simply 2.3 kilo-meters. The power is chosen so that the final number will be between 1 and 999.

Example: Express the number 64,000 in engineering notation.

$$64,000 = 64 \times 10^3$$

Example: Express the number 0.0000456 in engineering notation.

$$0.0000456 = 45.6 \times 10^{-6}$$

REVIEW PROBLEMS

Express the following values in scientific notation.

1. A Firewire port supports data speeds up to 3,200,000,000 bps.   _____

2. The sampling rate of a CD is 44,100 Hz.   _____

3. A Pentium IV has 8,192 bytes of L1 cache for data.   _____

4. The distance between transistors in a Core Duo processor is only 0.0000000065 m.   _____

5. A monitor has a refresh rate of 0.0125 s.   _____

Express the following values in engineering notation.

6. PC 6400 RAM runs at 800,000,000 Hz.   _____

7. A SATA hard drive can transfer data at 150,000,000 B/s.   _____

8. A T1 line sends data at a rate of 1,544,000 bps.   _____

9. The core of a piece of single-mode fiber cable has a diameter of 0.000008 m.   _____

10. A hard drive has a 0.013 s access time.   _____

Express the following values in normal (decimal) notation.

11. In 2006, there were approximately 3.95×10^8 hosts on the Internet.   _____

12. Some types of motherboard BIOS can only support hard drive partitions up to 1.37×10^{11} bytes.   _____

13. Your motherboard supports a 1.066×10^9-Hz frontside bus.   _____

14. A benchmark program determines that it takes your computer 2.4×10^{-9} seconds to add two integers.   _____

15. PC-133 RAM has a refresh rate of 7.5×10^{-8} seconds.   _____

16. The Ping command measures an average response time of 23×10^{-3} s.   _____

17. A single-mode fiber cable uses laser light at a wavelength of 1550×10^{-9} m.   _____

18. A hard drive contains 500×10^{9} bytes of data.   _____

19. A blank CD-R can hold 700×10^{6} bytes of data.   _____

20. The speed of light in a fiber is 2.1×10^{8} m/s.   _____

Unit 21 SIGNIFICANT DIGITS AND ROUNDING

BASIC PRINCIPLES

The accuracy of a measurement is not always clear. If someone tells you they measured 150 feet of network cable, you don't know if this figure is accurate to the nearest 10 feet or the nearest foot. To judge the accuracy of a measurement you need to determine which values were actually measured and which are merely place holders. The measured numbers are *significant* to the accuracy of the answer.

All numbers are significant—except for the zeros whose only purpose is to set the decimal place. For example, the zeros in 0.00035 and 2,400 are not significant so both of these numbers have only two significant figures.

The zeros in 1,002 and 35.00 *are* significant because they do more than mark the decimal place; so both of these numbers have 4 significant figures. The addition of a decimal point to the right of a number is often a clue that some of the zeros are significant. The number 100 only has one significant figure. If expressed as 100.0, however, four figures are significant.

Example: How many significant figures are in the number 123,000?

As it is written, there are three significant figures (1, 2, and 3). In order to express this number with more significant figures it would have to be written in scientific notation.

The number 1.23×10^5 has three significant figures, while 1.2300×10^5 has five significant figures.

ROUNDING

When several measurements are combined, the answer sometimes contains a greater number of significant figures than you can be sure of. The answer should only have as many significant figures as the least accurate measurement. If the answer has too many significant figures, a process called *rounding* is used to reduce the number of significant figures in the answer.

Use the following rules to round a number:

1. Locate the number to be rounded (the least significant figure) in your answer by counting the required number of significant figures from the left. Remember, your answer should have the same number of significant figures as the least accurate measurement.

2. If the number to the right is 5 or more, increase the value of the least significant figure by one.

3. Leave the least significant figure alone if the number to the right is less than 5.

Example: Round the number 1.674278 to five significant figures.

The least significant figure is 2. The number to the right of 2 is 7, which is greater than 5, so the least significant figure (2) can be "rounded up" to 3. The answer is 1.6743.

Example: Add the numbers 4.1 and 3.214 and then round the answer to the correct number of significant figures.

> 4.1 (two significant figures) + 3.214 (four significant figures)
> = 7.314 (four significant figures)

Since one of the numbers being added has only two significant figures, the answer must be rounded to only two significant figures as well.

> 7.314 rounds to 7.3 (two significant figures)

REVIEW PROBLEMS

Determine how many significant figures are contained in the following measurements and calculations.

1. 321 feet of Cat 6 cable.   _____

2. 327.08-GB hard drive.   _____

3. 420,000 files.   _____

4. 34.00-KB file.   _____

5. 0.0245-GB folder.   _____

Solve the following problems and round the answers to the correct number of significant figures.

6. The sum of a 56-GB volume and a 16.7-GB volume.   _____

7. The total length of the following cables: 300.0 feet, 43.45 feet, 57 feet.   _____

8. Your computer is running four processes. They use 17 KB, 26.6 KB, 120.1 KB, and 28.88 KB. Find the sum.   _____

9. A hard drive is 120.00 MB large but contains 78.3 MB of files. How much space is left on the drive?   _____

10. You decide to time your computer and find it takes 45 s to boot. If a boot optimizer will reduce this time by 6.83 s, how long will it take to boot?   _____

11. A motherboard has a 533-MHz frequency and runs at a 4.5× multiplier. Find the product to determine the CPU speed.   _____

12. There are 0.3048 meters in a foot. How many meters long is a 12-foot cable?   _____

13. You measure a monitor as 20.1 inches by 15.86 inches. What is the surface area of the monitor?   _____

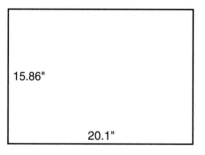

14. You determine that hard drives currently cost $0.8 per GB. How much would you expect to pay for a 565-GB hard drive?   _____

15. Video card A can render 35 fps and video card B can render 180.56 fps. How much faster is video card B?   _____

Round the following values as instructed.

16. A CD player requires 25.52 watts. Round this value to the nearest watt.   _____

17. A hard drive contains 564.9 GB of data. Round this value to the nearest 10 GB.   _____

18. A display has a refresh rate of 0.001578 s. Round this value to nearest hundredth of a second.   _____

19. A software license cost $99.50. Round this value to the nearest dollar.   _____

20. A T1 line has a data transfer rate of 1,544,000 bps. Round this figure to the nearest 100,000 bps.   _____

 # Unit 22 *EQUIVALENT UNITS*

BASIC PRINCIPLES

Most scientists use engineering notation and the metric system for recording measurements. (The metric system is also called the International System of Units or SI.) To illustrate, rather than write 2,300 Hz, most scientists would write 2.3×10^3 Hz or simply 2.3 kHz. The following table lists some of the most common SI prefixes and their symbols. Notice how the exponent increases by 3 as you move up the table.

Prefix	Symbol	Exponent
peta	P	10^{15}
tera	T	10^{12}
giga	G	10^{9}
mega	M	10^{6}
kilo	k	10^{3}
milli	m	10^{-3}
micro	μ	10^{-6}
nano	n	10^{-9}
pico	p	10^{-12}
femto	f	10^{-15}

In the field of computer science, there is one exception to this rule: Kilobyte is ambiguous. Sometimes it means $2^{10} = 1,024$ bytes; other times it means 1,000 bytes—often in the same documentation! This difference can cause some confusion when determining the size of a hard drive since different diagnostic programs will often use different senses of kilobyte.

Example: Express the value 3.45×10^9 bps using SI prefixes.

3.45×10^9 bps = 3.45 Gbps

Example: Express the value 1,234 MHz as GHz.

1,234 MHz = 1.234 GHz

REVIEW PROBLEMS

1. The voltage of the CMOS battery is measured at 2,896 millivolts. Express this measurement in volts.   _____

2. You purchase a 700-gigabyte hard drive. Express this value in terabytes.   _____

3. You have a 12,463,287-byte file. Express its size in megabytes.   _____

4. The frontside bus runs at 533 MHz. Determine this value in Hz.   _____

5. The data transfer rate over a network is measured at 4.5 Mbps. What would be the rate in Kbps?   _____

6. Wireless signals move at 300,000,000 m/s. Express this value in km/s.   _____

7. PC-100 RAM has a refresh rate of 10 nanoseconds. Express this value in milliseconds.   _____

8. USB 2.0 can directly power devices that draw up to 1 amp. Express this value in milliamps.   _____

9. A transistor on a CPU is 45 nm across. What is its size in millimeters?   _____

10. The core in a piece multimode fiber has a diameter of 62.5 μm. Express this diameter in millimeters.   _____

11. A single-layer DVD can hold 4.7 GB of data. Express this value in MB.   _____

12. Coaxial Ethernet cable uses a 50 Ω (ohm) terminator. Express this resistance in kΩ.   _____

13. One of the capacitors in a computer's switching power supply measures 100 μF. Express the capacitance in mF.   _____

14. Bluetooth devices run in the 2.4-GHz range. Express this value in MHz.   _____

15. A new video card claims to have 4 teraflops of computing power. Express this value in megaflops.   _____

16. USB 2.0 supports speeds of up to 480 Mbps. Express this value in Kbps.   _____

17. A fiber optic cable runs 54.8 km. Express this distance in meters.   _____

18. Windows XP can accurately measure a time delay to about 10 ms. Express this time in seconds.   _____

19. The cycle time for a 1-GHz processor is 1 nanosecond. Express this time in microseconds.   _____

20. By the year 2030, most computers will run in the 1-PHz range. Express this frequency in GHz.   _____

Formulas and Symbols

Unit 23 FUNCTIONS AND FORMULAS

BASIC PRINCIPLES

In mathematics, a *formula* represents equality between two sets of quantities that can be expressed symbolically as an *equation*. Programmers often use a kind of formula called a *function* to transform specific inputs into output values.

An equation groups various numbers, letters, and other symbols on either side of an equal sign (=) to show a relationship between properties represented by symbols called variables. By substituting known values for all but one of these variables we can *solve* the equation.

Example: Find the area of a rectangle that measures 4 feet by 3 feet.

The formula for the area of a rectangle is $A = lw$ where "A" stands for area, "l" stands for length, and "w" represents width. If we substitute numerical values for l and w we can solve the equation and find a value for A.

$$\text{Area} = 4 \text{ feet} \times 3 \text{ feet} = 12 \text{ square feet}$$

The values in a formula can sometimes be rearranged by multiplying or dividing both sides of the formula by a common variable. The formula $A = B \times C$ could also be written as $B = \frac{A}{C}$ or $C = \frac{A}{B}$ or by dividing the whole equation by either B or C.

Example: Express the formula $a = \dfrac{F}{m}$ in terms of F.

$$a = \frac{F}{m} \qquad m \times a = \frac{m \times F}{m} \qquad m \times a = F \qquad F = m \times a$$

In this case we multiplied both sides by the variable *m* to leave *F* alone on one side of the equals sign.

REVIEW PROBLEMS

1. A Web designer needs to calculate the area of a rectangle on the screen. The formula for area is $A = lw$, where A = area, l = length, and w = width. What is the area (A) when l = 5 and w = 2?   _____

2. The formula for the area of a triangle is $A = \frac{1}{2}bh$, where A = area, b = base, and h = height. How many pixels would be in a triangular graphic where b = 20 pixels and h = 30 pixels?   _____

3. A network technician needs to run a cable around a circular table. The formula for the circumference of a circle is $C = \pi d$, where $\pi = 3.14$ and d = diameter. How long is the cable if d = 5 feet?   _____

4. A computer technician needs to calculate the volume of a computer case in order to determine whether or not a larger case fan is required. The formula for the volume of a rectangular solid is $V = lwh$, where V = volume, l = length, w = width, and h = height. What is the volume in cubic inches if the case has a length of 20 inches, a width of 15 inches, and a height of 5 inches?   _____

5. Calculate the watts used by a 12-V case fan that runs at 2 A. The formula for watts is $W = VA$, where W = watts, V = volts, and A = amps.   _____

6. The BIOS reports the CPU temperature as 45 degrees Celsius. The formula for converting from Celsius to Fahrenheit is $F = (C \times 1.8) + 32$, where F = degrees Fahrenheit and C = degrees Celsius. What is the CPU temperature in degrees Fahrenheit?   _____

7. A call center manager is trying to estimate the cost of electricity to operate a computer for one week. The formula for calculating the cost is $C = (Wtc)/1{,}000$, where C = the cost in cents, W = the average watts used by the computer, t = the time in hours, and c = cost per kilowatt hour. What is the cost if W = 350 watts, t = 168 hours, and c = $0.10?   _____

8. A programmer needs to check the results of the equation $Z = 3x + 2y$. Calculate the value of Z if x = 6 and y = 5.   _____

9. The formula for calculating the actual data transfer rate of a hard drive is $R = d/t$, where R = the data transfer rate, d = the amount of data transferred, and t = the time it took to transfer the data in seconds. What is the data transfer rate if d = 80 MB and t = 25 seconds?

    _____

10. Modern graphics are often comprised of many polygons. Calculate the area of the polygon below using the formula $A = (a + b)h/2$.

    _____

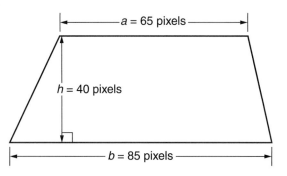

    $a = 65$ pixels

    $h = 40$ pixels

    $b = 85$ pixels

11. The memory required to save an uncompressed image file can be represented with the formula $M = (tb)/1{,}024$, where M = the memory required in KB, t = the total number of pixels in the image, and b = bytes used to store the color of each pixel. Find the memory required to store a 640-pixel by 480-pixel image where b = 3 bytes.

    _____

12. A rectangular frame on a Web page has an area of 80,000 pixels and a width of 200 pixels. If the formula for area is $A = lw$, where A = area, l = length, and w = width, what is the length of the frame?

    _____

13. You measure the circumference of a CD to be 376.8 mm. The formula for the circumference of a circle is $C = \pi d$, where $\pi = 3.14$ and d = diameter. Solve for d.

    _____

14. The size of an uncompressed audio file can be calculated with the formula $S = fat$, where S = the size of the file, f = the frequency in Hz, a = the accuracy of the sample in bytes, and t = the time of the recording in seconds. A file of 352,800 bytes, was recorded at a frequency of 44,100 Hz, and a 2-byte sample accuracy. What is the length of the recording?

    _____

15. The physics engine in a video game uses the formula $F = m \times a$ to calculate the force of a moving object. Solve for mass (kg) if $F = 50 \left(kg \times \frac{m}{s^2}\right)$ and acceleration $= 10 \left(\frac{m}{s^2}\right)$.

    _____

16. A program contains the function A = 2 (b + c) + 3. If A = 19 and b = 2, what is the value of c?   _____

17. A program to convert pounds to kilograms uses the formula $kg = lb \times 0.454$. What formula would you use to convert kilograms to pounds?   _____

18. Express the width of the following table in terms of the size of its cells.   _____

A1	B1	C1
A2	B2	C2
A3	B3	C3
A4	B4	C4

19. A service technician charges $40 up front plus $25/hr for house calls. Express this rate as an equation.   _____

20. A wireless Internet service provider charges a flat $15 monthly fee plus $10/hr. How many hours would you have to be connected to get a monthly bill of $195?   _____

Unit 24 EXPONENTIAL AND LOGARITHMIC FUNCTIONS

BASIC PRINCIPLES

Some equations may also include powers, roots, or logarithms. Section 5 introduced powers (exponents). A *root* is the reverse of a power. An exponent represents the number of times a number must be multiplied by itself, whereas a root determines what number must be multiplied by itself to get a given product.

Example: What is the square root of 25?

Finding the square root of 25 involves finding a number that can be squared to give a product of 25. This number is 5.

$$\sqrt{25} = 5$$

Notice this is the reverse of a power of 2.

$$5^2 = 25$$

Logarithms (logs) are also related to powers. Suppose you wanted to know what power 10 must be raised to in order to yield 100? The answer is 2 because $10^2 = 10 \times 10 = 100$. This would be expressed as $\log_{10} 100 = 2$. Not all logs work out to be whole numbers. For complicated logs and roots it is necessary to use a calculator.

Example: Find $\log_{10} 75$ and round the answer to 2 decimal places.

$$\log_{10} 75 = 1.875061 \ldots \approx 1.88 \text{ (approximately equal to 1.88)}$$

REVIEW PROBLEMS

1. To calculate the areal density (bits per square inch) of a hard drive, you first must calculate the surface area of the platters. Calculate the surface area of one side of a hard drive platter with the formula *Area* = πr^2, where π = 3.14 and r is the radius of the drive. The drive has a radius of 2.5 inches.

_____

2. An optical fiber has a core diameter of 62.5 microns. If the radius is half the diameter, what is the cross-sectional area of the cable to the nearest 10 microns?

_____

3. A computer speaker has a power output of 12 watts and a nominal impedance of 4 ohms. How many volts is it using, given the formula *volts* = $\sqrt{watts \times ohms}$? Round the answer to the nearest tenth of a volt.

_____

4. The water from a liquid-cooled CPU drains into the cylinder depicted below. How much liquid will it hold, given the formula *volume* = $\pi r^2 h$, where π = 3.14, r = the radius, and h = the height.

_____

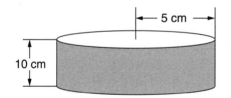

5. What is the volume of a cubic computer case that measures 12 inches in each direction? The formula for the volume of a cube is $V = s^3$, where s = the length of one side.

_____

6. A Java program contains the function $y = 3^x - 2x$. Solve for y, assuming that x has a value of 4.

_____

7. A game programmer needs to calculate the volume of a sphere with the function $V = \frac{4}{3}\pi r^3$. Solve for V if r has a value of 2 feet.

_____

8. The current requirements of a case fan can be expressed by the equation *current* = $\sqrt{\frac{watts}{ohms}}$. Find the current for a 5-watt case fan that has a resistance of 8 Ω (ohms).

_____

9. A lab technician needs to calculate the light level at her lab bench. The formula for calculating light level is *foot-candles* $= \frac{candelas}{distance^2}$. How many foot-candles would you expect from a 5,000-candela light source at a distance of 8 feet?

    _____

10. The area of a square display is 289 square inches. The length can be calculated with the function $l = \sqrt{area}$, where l = the length of one side. Find the length of one side.

    _____

11. You need to calculate the area of a CD label. Use the formula $A = \pi(R^2 - r^2)$, where A = area, π = 3.14, R = the outer radius, and r = the inner radius.

    _____

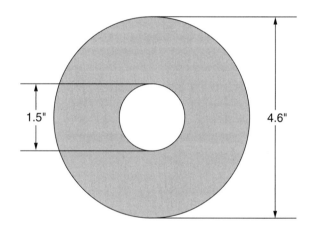

12. A programmer is using the equation $y = 3x^2 + 2x + 4$ to write a program that will estimate the stopping distance for different kinds of tires. Solve for y if x = 3.

    _____

13. Find the diagonal length (to the nearest inch) of an LCD display that measures 15 inches by 27 inches. The formula for diagonal length is $z = \sqrt{x^2 + y^2}$.

    _____

14. Use the formula for diagonal length to find the cable run distance from opposite sides in a room that measures 30 feet by 40 feet.

    _____

15. A wireless router claims to cover a circular area of 3,000 square meters. What is the maximum distance that you can connect to the router? Use the formula *Area* $= \pi r^2$, where π = 3.14 and r is the radius. Round the answer to the nearest meter.

    _____

16. The value of a computer halves every year according to the function *Value = Initial Price* $\times$ $(\frac{1}{2})^N$, where N = the number of years since the computer was purchased. What would be the value of a 3-year-old computer that was purchased for $2,000? Round the answer to the nearest dollar.

       _____

17. An astronomy program uses the equation $y = \log_{10} x$ when plotting the brightness of some stars. What would be the value of y if x had a value of 100?

       _____

18. A programmer needs to check the output of the function $y = 2x^2 + x \log_{10} 100$ where x has a value of 5. Solve this equation.

       _____

19. The minimum number of steps it will take to sort a list of numbers can be expressed with the function $t = n(\frac{\log_{10} n}{\log_{10} 2})$. Solve for t if n = 20.

       _____

20. The signal loss in a network cable is measured in decibels with the formula $dB = 20 \log_{10}(\frac{V_1}{V_2})$, where V_1 is the voltage at the end of the cable and V_2 is the initial voltage. What would be the signal loss for a voltage that drops from 1 V to 0.5 V?

       _____

Unit 25 REARRANGING FORMULAS

BASIC PRINCIPLES

Formulas can be manipulated as long as both sides of the equation remain equal. An equation can often be simplified by adding or subtracting the same value to both sides of an equation.

Example: Solve the equation $5 + x = 8$.

$5 + x = 8$	(original equation)
$5 + x - 5 = 8 - 5$	(subtract 5 from both sides)
$x = 3$	(solution)

The same technique can also be used for multiplication and division, as well as powers and roots. In all three cases (addition and subtraction, multiplication and division, and powers and roots) the application of one operation cancels the other.

Example: Solve the equation $\sqrt{x} = 3$.

$\sqrt{x} = 3$	(original equation)
$(\sqrt{x})^2 = (3)^2$	(raise both sides to a power of 2)
$x = 9$	(solution)

REVIEW PROBLEMS

Programmers often have to work with formulas to do everything from plotting graphics to calculating interest. Rearrange the following formulas for the unknown value.

1. $2x = 10$ _____

2. $\dfrac{y}{3} = 12$ _____

3. $16 + b = 45$ _____

4. $\dfrac{x+2}{3} = 15$ _____

5. $2b + 3b = b + 8$ _____

6. $5 + \dfrac{x}{1} = 10$ _____

7. $7 - n = 7$ _____

8. $1.2 + a - 0.3 = 3.4$ _____

9. $25 = \overline{)x - 1}$ _____

10. $7m - 1 = 15 - m$ _____

Often questions will have more than one variable. Solve for the indicated variable.

11. $c = \pi d$ Solve for d. _____

12. $A = \dfrac{1}{2}bh$ Solve for b. _____

13. $y = mx + b$ Solve for x. _____

14. $area = \pi ab$ Solve for a. _____

15. $y = mx + b$ Solve for b. _____

16. $°F = (°C \times \dfrac{9}{5}) + 32°$ Solve for °C. _____

17. $watts = volts \times amps$ Solve for volts. _____

18. $area = \pi \times radius^2$ Solve for radius. _____

19. $z = \sqrt{x^2 + y^2}$ Solve for x. _____

20. $volume = \pi r^2 h$ Solve for r. _____

Numbering Systems

 **Unit 26 BINARY REPRESENTATION**

BASIC PRINCIPLES

Binary (also called base 2) representation is a numbering system that uses only two symbols, 0 and 1. Understanding binary is necessary in information technology since computers represent data in binary form. Each *binary digit,* or bit, has a different value depending upon where it is located in a binary number. The least significant digit, on the right, has a value of 1, and each bit to the left is worth twice as much as the previous one.

Example: Find the decimal equivalent of the binary number 1011.

To convert this number to decimal form, the value of each bit must first be calculated.

Binary Digit	1	0	1	1
Place Value	8	4	2	1

The binary number 1011 means:

$$(1 \times 8) + (0 \times 4) + (1 \times 2) + (1 \times 1) = 8 + 2 + 1 = 11$$

Example: Find the binary equivalent of the decimal number 135.

Start by listing all of the powers of two that are less than 135. Note that as you move to the left, each place value is worth twice as much as the previous one.

2^7	2^6	2^5	2^4	2^3	2^2	2^1	2^0
128	64	32	16	8	4	2	1

The number 135 contains one 128 (2^7) and leaves a remainder of 7. The values 64, 32, 16, and 8 are not used in representing 135; but the 4, 2, and 1 are used to represent the remainder of 7. Using the binary system, there is only one combination of ones and zeros that will add to exactly 135. The decimal number 135 means $128 + 4 + 2 + 1$ or 10000111 in binary.

1	0	0	0	0	1	1	1
128	0	0	0	0	4	2	1

Example: Find the binary equivalent of the decimal number 31.

The number 31 contains one 16 (2^4), one 8 (2^3), one 4 (2^2), one 2 (2^1), and one 1 (2^0). This is represented with the binary number 11111. Note that this is only one digit less than 32_{10} (decimal) which is 100000_2 (binary).

REVIEW PROBLEMS

1. The preamble before a data communication is often written as a binary series of ones and zeros. Convert the binary number 10101010 into decimal.   _____

2. A byte is made up of eight bits. What is the decimal value of the largest number that can be represented with one byte?   _____

3. Write out all of the possible combinations of 0s and 1s contained in three bits. Place them in ascending order from the smallest to the largest.   _____

4. A nibble is made of four bits. What is the decimal range of all nibbles?   _____

5. Convert the IP address 10.13.110.209 into four 8-bit binary numbers.   _____

6. Convert the subnet mask 255.240.0.0 into four 8-bit binary numbers.   _____

7. Class "A" IP addresses all start with a zero in their first octet. What are the decimal equivalents of the highest and lowest 8-bit binary numbers that start with a zero?   _____

8. Class "C" IP addresses all start with 110 in their first octet. What is the decimal range of all class "C" addresses?
    _____

9. A valid subnet mask changes from ones to zeros only once. Convert the subnet mask 255.255.192.0 into four 8-digit binary numbers to determine if it is valid or not.
    _____

10. Convert the subnet mask 11111111.11111111.11100000.00000000 into four decimal numbers.
    _____

11. Convert the subnet mask 11111111.11111100.00000000.00000000 into four decimal numbers.
    _____

12. Many SCSI devices use four jumpers to set the SCSI ID as a 4-digit binary number. Open jumpers represent zero, while closed jumpers mean one. How would you set the SCSI ID to a value of 5?
    _____

13. Suppose you have a network card that uses dip switches to set its IRQ. The switches are currently set to "on, off, on, on" where "on" represents 1 and "off" represents 0. If the least significant digit is on the right, what is the IRQ?
    _____

14. ASCII code represents the symbol "?" with the decimal number 63. Convert this number to an 8-bit binary number.
    _____

15. ASCII code represents the symbol "!" with the binary number 00100001. Convert this number to decimal form.
    _____

16. Linux file permissions are set by a 3-digit binary number where the three bits represent the read, write, and execute permissions. A 1 means the permission is enabled and a 0 means it is disabled. What permissions would be represented by a decimal value of 5?
    _____

17. You need to store a value that ranges from 0 to 63. How many bits are required to store this number?
    _____

18. How would you represent 1 K (1,024) as a binary number?
    _____

19. A single byte is used to define the characteristics of each file on a hard drive according to the following table. Would a file with an attribute byte value of 7 have the hidden attribute? _____

File Attribute	Attribute Bit
Read-Only	00000001
Hidden	00000010
System	00000100
Volume Label	00001000
Directory	00010000
Archive	00100000

20. The Microsoft Windows XP registry specifies when to disable the AutoRun feature according to the following table. The default setting is decimal number 149, which disables AutoRun on unknown, removable, and network devices. How should this decimal number be changed to disable AutoRun on CD-ROMs as well? _____

Drive Type	Control Bit
Unknown Device	00000001
No Root Directory	00000010
Removable Drive	00000100
Hard Disk	00001000
Network Drive	00010000
CD-ROM	00100000
RAM Disk	01000000
Reserved	10000000

Unit 27 BINARY ADDITION AND MULTIPLICATION

BASIC PRINCIPLES

The processes of adding and multiplying binary numbers are the same as with decimal numbers. Numbers are added from left to right and carried when the sum is more than one digit. Multiplying two binary numbers involves adding a series of products that are shifted for their place value.

Example: What is the sum of the binary numbers 101 and 10?

Start by adding the numbers in the right hand column. They are a 1 and a 0, so their sum is 1. The next column contains a 0 and a 1, which also equals 1. The third column only contains a 1, so its total is also 1.

$$
\begin{array}{cc}
101 \\
+\,10 \\
\hline
\end{array}
\rightarrow
\begin{array}{cc}
101 \\
+\,10 \\
\hline
1
\end{array}
\rightarrow
\begin{array}{cc}
101 \\
+\,10 \\
\hline
11
\end{array}
\rightarrow
\begin{array}{cc}
101 \\
+\,10 \\
\hline
111
\end{array}
$$

The accuracy of your answer can be checked by converting all of the numbers to decimal values.

101 (5 in decimal) + 10 (2 in decimal) = 111 (7 in decimal)

Example: What is the sum of the binary numbers 1011 and 11?

The right-hand column contains two 1s which add to 10 (2 in decimal). Write down the 0, and carry the 1 to the next column. The second column (to the left) now contains three 1s, which add to 11 (3 in decimal). Write down a 1 in this column, and carry the other 1 to the next column to the left. Continue this process to get a sum of 1110.

$$
\begin{array}{cc}
1011 \\
+\,11 \\
\hline
\end{array}
\rightarrow
\begin{array}{cc}
1 \\
1011 \\
+\,11 \\
\hline
0
\end{array}
\rightarrow
\begin{array}{cc}
1 \\
1011 \\
+\,11 \\
\hline
10
\end{array}
\rightarrow
\begin{array}{cc}
1011 \\
+\,11 \\
\hline
110
\end{array}
\rightarrow
\begin{array}{cc}
1011 \\
+\,11 \\
\hline
1110
\end{array}
$$

1011 (11 in decimal) + 11 (3 in decimal) = 1110 (14 in decimal)

Example: What is the product of the binary numbers 101 and 11?

The product of 101 and 11 is equivalent to the sum of $101 \times 1 = 101$ and $= 101 \times 10 = 1010$.

101		101		101		101
× 11		× 11		× 11		× 11
	→	101	→	101	→	101
				1010		1010
						1111

101 (5 in decimal) + 11 (3 in decimal) = 1111 (15 in decimal)

As mentioned in the previous unit, the number of unique values a binary number can hold is calculated by multiplying 2 by itself once for each binary digit. This means that a 2-digit binary number can have 4 four different values: 00, 01, 10, and 11.

Example: How many values can be represented with a 6-digit binary number?

There are $2 \times 2 \times 2 \times 2 \times 2 \times 2 = 2^6 = 64$ possible combinations of six 0s and six 1s. They range from 000000 to 111111.

REVIEW PROBLEMS

1. Find the sum of the binary numbers 100 and 011.   _____

2. Find the sum of the binary numbers 1010 and 1101.   _____

3. Find the product of the binary numbers 100 and 111.   _____

4. Find the product of the binary numbers 0111 and 1010.   _____

5. Find the sum of the byte numbers 00011001 and 10110011.   _____

6. Find the sum of the byte numbers 11111111 and 00000001.   _____

7. Find the product of the byte numbers 00010001 and 00110011.   _____

8. Find the product of the byte numbers 10111011 and 01111101.   _____

9. Calculate the output of a 4-bit full adder circuit with inputs of 1001 and 0111.   _____

10. Determine whether or not the sum of the 8-bit numbers 11010011 and 00110000 would cause an overflow error by exceeding the largest possible 8-bit number.   _____

11. Determine the minimum number of bits required to store the sum of the binary numbers 1111 and 1111.   _____

12. Determine the minimum number of bits required to store the output of eight different 4-bit numbers.   _____

13. What is the largest decimal value that can be stored with one nibble (4 bits)?   _____

14. What is the largest decimal value that can be stored with one byte (8 bits)?   _____

15. How many colors can be displayed if a video card uses 24 bits to store each color?   _____

16. If a computer added 4 lines (bits) to its memory address bus, how many times more memory could it hold?   _____

17. A CD stores audio by recording the voltage many times each second. How many unique voltages are recorded with 16-bit audio?   _____

18. Computers usually represent text with a 7-bit encoding scheme called ASCII. How many different characters can be represented with 7 bits?   _____

19. The extended ASCII set has 8-bits per character instead of the 7 used in regular ASCII. How many new characters does the extended ASCII set add?   _____

20. A hard drive formatted with FAT32 has 32 bits per address. How many unique allocation units can be addressed on a FAT32 hard drive?   _____

Unit 28 *HEXADECIMAL AND OCTAL NUMBERING*

BASIC PRINCIPLES

Hexadecimal numbering (also called hex or base 16) uses 16 symbols (for counting from 0 to 9 and from A through F). The least significant digit represents the number of 1s, with every digit to the left being worth 16 times more. The same pattern can be followed with *octal numbering* (base 8).

The key to counting with hex is to remember not to stop at 9 but to continue on with A through F before carrying the 1.

Example: Count the first 20 decimal numbers in hex.

0, 1, 2, 3, 4, 5, 6, 7, 8, 9, A, B, C, D, E, F, 10, 11, 12, 13, 14

Dec	0	1	2	3	4	5	6	7	8	9	10	11	12	13	14	15	16	17	18	19	20
Hex	0	1	2	3	4	5	6	7	8	9	A	B	C	D	E	F	10	11	12	13	14

The number 14 (one four, not fourteen) in hex means 20 in decimal because it contains one 16 and four 1s.

Example: Find the decimal equivalent of the hex number 2C.

To convert this number to decimal, the value of each bit must first be calculated.

Hex Digit	2	C
Place Value	16	1

The hex number 2C means $(2 \times 16) + (12 \times 1) = 32 + 12 = 44$

Example: Find the hex equivalent of the decimal number 135.

Start by listing all of the powers of 16 that are less than 135.

16^1	16^0
16	1

135 contains eight 16s, and leaves a remainder of 7.

Thus, the decimal number 135 means $128 + 7 = (8 \times 16) + (7 \times 1) = 87$ in hex.

Example: Find the octal equivalent of the decimal number 135.

Start by listing all of the powers of 8 that are less than 135.

8^2	8^1	8^0
64	8	1

135 contains two 64s, and leaves a remainder of 7. There are no 8s, but there are seven 1s.

Therefore, decimal number 135 means $128 + 0 + 7 = (2 \times 64) + (0 \times 8) + (7 \times 1)$ or 207 in octal.

The number of unique values a hex number can hold can be calculated by multiplying 16 by itself once for each digit. Thus, a 2-digit hex number can have 256 values ranging from 00 to FF.

Example: How many values can be represented with a 4-digit hex number?

There are $16 \times 16 \times 16 \times 16 = 16^4 = 65{,}536$ possible combinations ranging from 0000 to FFFF.

REVIEW PROBLEMS

1. Computers use hex numbers to assign resources such as interrupt request channels (IRQs) to devices. Convert IRQ 15 into a hex number.   _____

2. Convert the hex number 20h (the ASCII character for a space) into decimal.   _____

3. COM1 uses the eight I/O addresses from 3F8h to 3FFh. Convert these numbers to decimal.   _____

4. A network analyzer was used to capture a destination IP address from an IP packet and displayed the results in hex form as 0A 0D 6F FA. Convert this number into a typical decimal IP address.   _____

5. You capture a 4-bit data stream from an I/O port of 0101. How would you express this number in hex?   _____

6. A POST diagnostic card is displaying the error 1Bh. Convert this to decimal so that you can look it up in a list of error codes.   _____

7. RGB color codes use three pairs of hex numbers to define a color. If all three pairs have the same value, the color is a shade of grey. Is the decimal color 4144959 grey?   _____

8. The decimal triplet (50, 205, 50) represents the color called lime green. Express this number as a hexadecimal triplet.   _____

9. How many bits are required to store a 2-digit hex number? **Hint:** How many bits are required to store the highest 2-digit hex number FFh?   _____

10. How would you express the value 2 KB (where KB = 1,024 B) in hex?   _____

11. Video BIOS starts at memory address C0000h and is 32 KB large. Find the sum to determine the address at the top of this memory region.   _____

12. How many bytes are required to store the product of two 3-digit hex numbers?   _____

13. The system BIOS uses memory addresses from F0000h to FFFFFh. How many bytes is this in decimal form?   _____

14. Web programmers sometimes use a 6-digit hex number to specify a color. How many colors can be defined using this method?   _____

15. Ethernet MAC addresses are 48-bits long. How many hex numbers are required to express a MAC address?   _____

16. The last six hex digits of a MAC address represent the serial number. How many different serial numbers are available?   _____

17. Convert the binary number 111 into octal.   _____

18. How many bits are required to store a 5-digit octal number?   _____

19. How would you represent 1 K as an octal number?   _____

20. Linux file permissions are set by a 3-digit binary number, where the 3 bits represent the read, write, and execute permissions. What octal number would you use to assign read and write permissions, but not execute?   _____

Unit 29 BCD AND ASCII ENCODING

BASIC PRINCIPLES

The ones and zeros that are stored on a computer are often meant to represent something other than a number. This is achieved through a process called encoding. Two common types of data that people often need to encode are decimal numbers and text files.

Although decimal numbers can be translated directly into binary, it is often more convenient to translate each digit separately in a type of encoding called Binary Coded Decimal or BCD. In BCD each decimal digit is represented by a 4-digit binary number. Since each digit is translated separately, it is much easier to check the answer for errors.

Example: Express the decimal number 123 in BCD.

Because 1 = 0001, 2 = 0010, and 3 = 0011, the whole number would be written as 000100100011.

Often the data in a computer is meant to represent text such as word-processing documents. The *American Standard Code for Information Interchange* (or ASCII) is used to represent symbols such as keyboard characters as well as some control characters. A table of ASCII characters can be found in the Appendix.

The symbol for a capital **A** can be found in the ASCII table (found in the Appendix) as decimal number 65. If this symbol were part of a word-processing document, it would appear in binary as 01000001.

Example: Express the ASCII string "abc" as a hex representation.

The symbols "abc" are decimal numbers 97, 98, and 99 in the ASCII character set. In hex, these numbers translate to 61h, 62h, and 63h.

REVIEW PROBLEMS

1. How many decimal digits can be stored in each byte using BCD? _____

2. You wish to store 9-digit telephone numbers in your database. Will this take up more space if they are saved as binary numbers or as BCD? _____

3. The computer's BIOS stores the date and time in BCD format. What month would have a value of 0101? _____

4. The computer's BIOS stores the date and time in BCD format. How would you encode September (the 9th month) in BCD? _____

5. You need to store a value that ranges from 0 to 127 in BCD format. How many bits are required to store this number? _____

6. In your database, you are converting a 24-bit binary field into BCD format. How many bits are required for the new field? _____

7. Convert the ASCII "@" symbol into BCD. _____

8. Convert the ASCII "@" symbol into binary. _____

9. What ASCII symbol corresponds to the BCD value 1000 0011? _____

10. What range of BCD numbers are unassigned in that they correspond to decimal values higher than 9? _____

11. Translate the ASCII string **Hello** into hex. _____

12. Translate the hex numbers 57 65 6C 6C 20 44 6F 6E 65 21 into ASCII. _____

13. Translate the ASCII string Hi into binary representation. _____

14. Translate the BDC code 01100110 10001001 01101001 into ASCII. _____

15. If you were writing a program to translate lowercase letters into uppercase letters, what decimal value would you have to subtract to each lowercase letter? _____

16. If you were writing a program to translate the ASCII symbols **0** through **9** into equivalent binary numbers, what 8-bit binary value would you have to subtract? _____

17. The ASCII form feed control character can be used to make a printer eject its current piece of paper. How would this character be expressed in binary? _____

18. Why do ASCII strings in Windows often group decimals 13 and 10 together?   _____

19. Web programmers will sometimes encode an e-mail address into another browser-compatible format to avoid spam. Turn the ASCII string first.last@work.com into a hexadecimal format.   _____

20. Bluetooth devices use a 16-character ASCII string as an authentication code for use with other devices. How many pairs of hex digits would be required to represent this string?   _____

Sets

Unit 30 SET THEORY

BASIC PRINCIPLES

Sets are simply collections of objects called *members* or *elements*. The members of a set can be any collection of distinct abstract objects like numbers or letters. IT professionals often use the mathematics of sets to work with databases—which are basically sets themselves. If a set is small, it can be described just by listing its members.

$A = \{a, b, c\}$

$B = \{110, 011, 001, 101\}$

$F = \{apples, oranges, bananas\}$

Large sets must be described by listing their properties using an ellipsis (three dots) to indicate the "understood" elements of a continuing pattern.

$N = \{1, 2, 3, \ldots\}$

$X = \{d, e, f, \ldots, q, r, s\}$

$E = \{n | n \text{ is a natural number}\}$

There are two sets of particular importance. The first is the set that contains no members $A = \{\ \}$, which is referred to as the *null set* or empty set and is represented with the symbol $\varnothing$. The null set is a member of any set even though it has no members itself. The second set of special importance is the set that contains all objects—the *universal set*, represented with the symbol $\cup$.

Example: Describe a set H that contains the hexadecimal digits.

$H = \{0, 1, 2, 3, 4, 5, 6, 7, 8, 9, a, b, c, d, e, f\}$

Often it is important to know exactly how many objects are in a set. For example, the manager of a database would need to know its size to allocate sufficient resources. This property of exactness is called *cardinality* and is usually represented by marking the set name with vertical lines on both sides.

Example: What is the cardinality of the set $A = \{a, b, c\}$?

Because A has 3 members, it is written as: $|A| = 3$.

REVIEW PROBLEMS

Use set notation to describe a set U with the following elements:

1. Windows XP, Windows Vista, Linux, and MAC OS.   _____

2. +5 V, –5 V, +3.3 V, 12 V, and –12 V.   _____

3. The binary numbers 00, 01, 10, and 11   _____

4. Whole numbers that are divisible by 2.   _____

5. All the members in the standard ASCII table from 00h to 7Fh.   _____

6. The universal set of all positive binary numbers.   _____

7. The set of all Windows operating systems created before 1960.   _____

Determine if each of the following is or is not a subset (or member) of the set of hexadecimal numbers from 0 to F.

8. 0   _____

9. A   _____

10. H   _____

11. $A = \{1, 2, 3, 4\}$ _____

12. $B = \{3, 6, 9, 10\}$ _____

13. $C = \{a, b, c, d, e\}$ _____

14.   D = {3, a, 6, c, f, 2}   _____

15.   E = { } or E = ∅   _____

Determine the cardinality for each of the following sets:

16.   A = {1, 2, 3, 4}   _____

17.   The set of binary values from 00 to 11.   _____

18.   The set of positive integers.   _____

19.   The set of ASCII characters in the word *Enter*.   _____

20.   B = ∅   _____

Unit 31 OPERATORS

BASIC PRINCIPLES

In earlier chapters we discussed some binary operators such as + (addition), − (subtraction), and × (multiplication) that could be used to combine binary digits in various ways. The comparable operators for sets include:

> union (A ∪ B), intersection (A ∩ B), and complement (∪-A or A').

Example: A = {1, 2, 3, 4, 5} and B = {3, 4, 5, 6, 7} What is the union of the sets A and B?

The *union* of these sets includes all of the members of A and all the members of B. Note, however, that where members are repeated, they only need to be written once.

> A ∪ B = {1, 2, 3, 4, 5, 6, 7}

Example: What is the intersection of above sets A and B?

The *intersection* of A and B includes only the members that appear in both A and B.

> A ∩ B = {3, 4, 5}

Example: What is the complement of A in B?

The *complement* of A in B includes all the members of B that are not in A.

> B − A = {6, 7}

Example: Let U = {a, b, c, d, e, f}, A = {a, b, d}, and B = {a, e, f} and solve (A ∩ B)'.

> A ∩ B = {a}

> (A ∩ B)' includes all the members of U except {a}; therefore, (A ∩ B)' = {b, c, d, e, f}

REVIEW PROBLEMS

Let U = {0, 1, 2, 3, 4, 5, 6, 7, 8, 9, a, b, c, d, e, f}, A = {2, 3, 6}, and B = {2, 5, 6, 7, a, d} and determine the following:

1. |A| _____

2. |B| _____

3. $|U|$ _____

4. $A \cup B$ _____

5. $A \cap B$ _____

Let U = {Carlos, Tom, Mary, Ahmed, Abby, Pat, Robin}, M = {Carlos, Tom, Ahmed, Pat, Robin}, F = {Mary, Abby, Pat, Robin} and determine the following:

6. M' _____

7. F' _____

8. $M \cup F$ _____

9. $M \cap F$ _____

10. $(M \cup F)'$ _____

Let U = {000, 001, 010, 011, 100, 101, 110, 111}, A = {000, 001, 100, 101}, B = {011, 101, 110} and determine the following:

11. $A \cup B$ _____

12. $A \cap B$ _____

13. $(A \cup B)'$ _____

14. $A \cup A$ _____

15. $|A \cup B| - |A \cap B|$ _____

Let U = {The set of even hex numbers}, A = {2, 4, 6}, B = {2, 4, 8}, C = {2, 6} and determine the following:

16. $|A \cap B \cap C|$ _____

17. $|A \cup B \cup C|$ _____

18. $|A| + |B| - |A \cap B|$ _____

19. $A \cup (B \cap C)$ _____

20. $A \cap (B \cup C)$ _____

Unit 32 VENN DIAGRAMS

BASIC PRINCIPLES

In set theory, a simple drawing called a *Venn diagram* is often used to represent a set operation. Venn diagrams use a rectangle to represent the universal set and circles to represent each of its subsets.

Example: Use a Venn diagram to represent $A \cup B$.

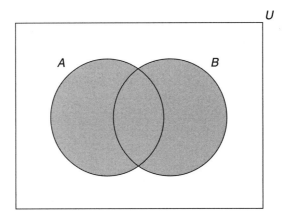

Example: Use a Venn diagram to represent $A \cap B$.

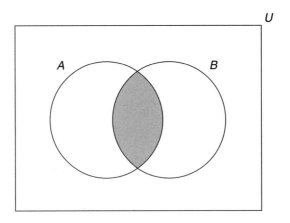

REVIEW PROBLEMS

Shade in the appropriate areas of the Venn diagrams to show the following operations:

1. *A'*

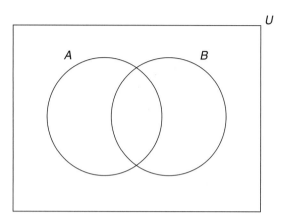

2. *A* ∩ *B'*

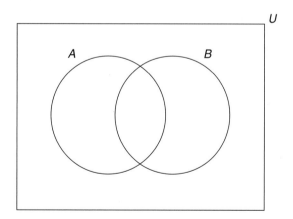

3. *B – A*

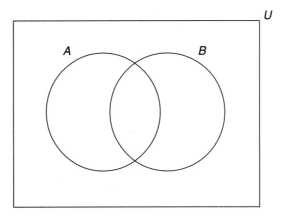

4. *(A ∪ B)'*

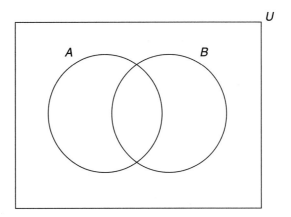

5.   *A'* ∪ *B'*

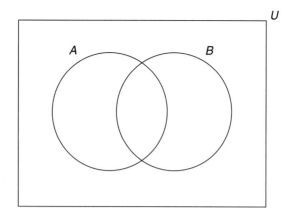

Use the sets U = {0, 1, 2, 3, 4, 5, 6, 7, 8, 9, a, b, c, d, e, f}, A = {1, 2, 3, a, b, c}, and B = {8, 9, a, b} to answer the following questions:

6.   Use a Venn diagram to illustrate sets U, A, and B.

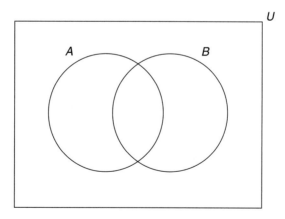

7. Demonstrate *B'*

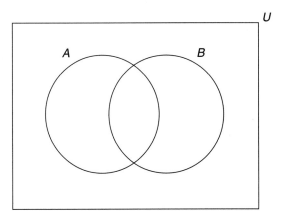

8. Demonstrate *A − B*

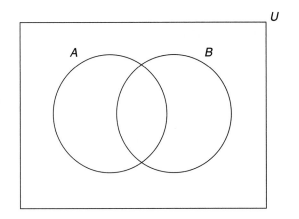

9.   Demonstrate $(A \cup B)'$

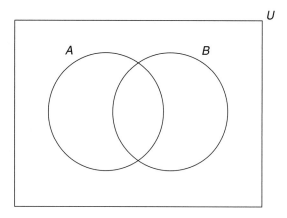

10.   Demonstrate $A' \cap B'$

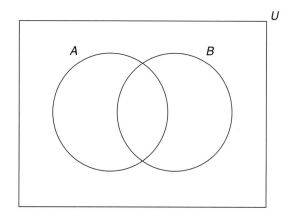

Use the sets U = {a, b, c, d, e, f, g}, A = {a, b, c}, B = {f, g}, C = {c, d} to answer the following questions:

11.   Use a Venn diagram to illustrate sets U, A, B, and C.

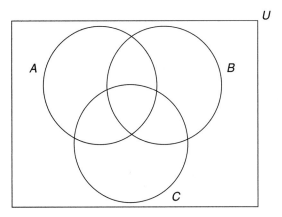

12.   Demonstrate *A'*

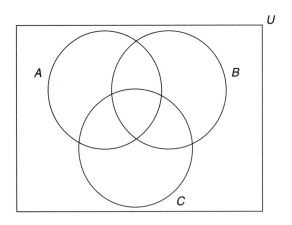

13.   Demonstrate $A \cap C$

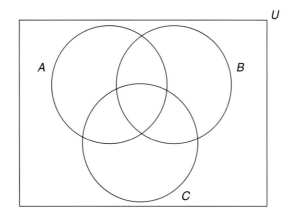

14.   Demonstrate $(A \cup B)'$

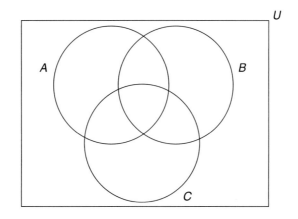

15.   Demonstrate $(A \cup B) \cap B$

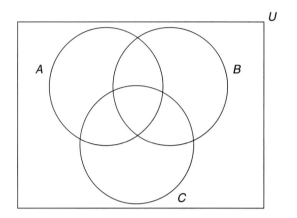

16.   Use a Venn diagram to illustrate the following cardinalities: $|A| = 2$, $|B| = 3$, $|A \cap B| = 0$.

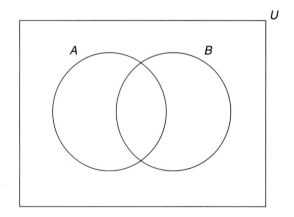

17. Use a Venn diagram to illustrate the following cardinalities: $|A| = 25$, $|B| = 15$, $|A \cap B| = 9$.

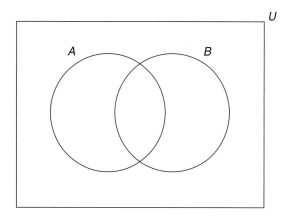

18. Use a Venn diagram to illustrate the following cardinalities: $|A| = 3$, $|B| = 4$, $|A \cup B| = 6$. **Hint:** $|A \cup B| = |A| + |B| - |A \cap B|$.

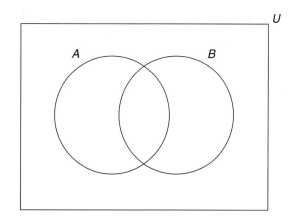

19. Use a Venn diagram to illustrate the following cardinalities: $|A| = 50$, $|B| = 75$, $|A \cup B| = 100$. **Hint:** $|A \cup B| = |A| + |B| - |A \cap B|$.

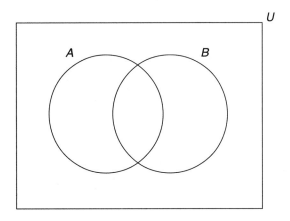

20. Use a Venn diagram to illustrate the following cardinalities: $|A| = 12$, $|B| = 12$, $|C| = 15$, $|A \cap B| = 3$, $|A \cap C| = 5$, $|B \cap C| = 4$, $|A \cup B \cup C| = 28$. **Hint:** $|A \cup B \cup C| = |A| + |B| + |C| - |A \cap B| - |A \cap C| - |B \cap C| + |A \cap B \cap C|$.

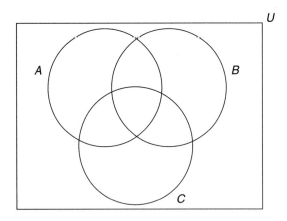

Unit 33 TRUTH TABLES

BASIC PRINCIPLES

Logic is basically the mathematics of making decisions. Logic connects input and output just like a computer does—in fact, both software and hardware are based on principles of logic. Although the input on a computer is usually in the form of ones and zeros, the input of a Boolean (logical) expression is made up of simple statements called *propositions*—and a proposition can be either true or false.

Not all sentences are propositions because not all sentences are either true or false. The statement "Today is Friday" is a proposition because it may be either true (when it is Friday) or false (when it is not Friday). Other sentences such as questions or orders are not propositions.

Example: Is the sentence "Please watch your step" a proposition?

No. Although it is a sentence, it does not make a statement that must be either true or false.

Just as sets may be combined with operators (such as ∪ or ∩) in logic, propositions also may be combined with operators. The three most common operators in logic are *AND*, *OR*, and *NOT*. The sentence, "My name is Todd, *AND* it is raining outside" is only true if it is also true that "My name is Todd" and "It is raining outside." *AND* is often represented with the ∧ symbol; *OR* is represented by ∨; and *NOT* is represented with ¬.

In order to understand how each of these operators affects the propositions they connect, we can make a table of all the possible inputs and outputs. This is referred to as a *truth table*.

Example: What is the truth table for the logical operator *AND*?

Input 1	Input 2	Output
p	q	$p \wedge q$
1	1	1
1	0	0
0	1	0
0	0	0

In this case, 1 stands for True, and 0 represents False. You can see that unless both input p AND q are true, the output is false.

The truth tables for OR and NOT are as follows:

p	q	$p \vee q$
1	1	1
1	0	1
0	1	1
0	0	0

p	$\neg p$
1	0
0	1

It is possible to combine more than one operator into a single expression.

Example: What is the truth table for the logical expression $\neg(p \vee q)$?

p	q	$p \vee q$	$\neg(p \vee q)$
1	1	1	0
1	0	1	0
0	1	1	0
0	0	0	1

REVIEW PROBLEMS

Determine whether or not each of the following sentences is a proposition.

1. Six plus seven equals eleven.   _____

2. Please pass the ketchup.   _____

3. Do you know the time?   _____

4. It's raining outside.   _____

5. My computer has two hard drives.   _____

Please complete the following truth tables.

6. $\neg p \vee q$

p	q	$\neg p \vee q$
1	1	
1	0	
0	1	
0	0	

7. $p \wedge \neg q$

p	q	$p \wedge \neg q$
1	1	
1	0	
0	1	
0	0	

8. $(p \vee q) \wedge \neg q$

p	q	$(p \vee q) \wedge \neg q$
1	1	
1	0	
0	1	
0	0	

9. $\neg p \wedge \neg q$

p	q	$\neg p \wedge \neg q$
1	1	
1	0	
0	1	
0	0	

10. $\neg(p \vee q) \wedge (\neg p \vee q)$

p	q	$\neg(p \vee q) \wedge (\neg p \vee q)$
1	1	
1	0	
0	1	
0	0	

For the following exercises, assume that propositions p and q are true and that propositions r and s are false. Determine if each expression is true or false.

11. $\neg p \vee q$ _____

12. $\neg(r \wedge s)$ _____

13. $\neg p \wedge (q \vee r)$ _____

14. $(p \wedge q) \vee (r \wedge s)$ _____

15. $\neg(r \wedge p) \vee (\neg s \vee q)$ _____

Here are some practical problems.

16. Most search engines support the logical operators *AND*, *OR*, and *NOT*.
     What command would you type if you wanted to search for cities called
     Paris that are not in France? _____

17. Write a command to search the Internet for crops grown in either North or South Carolina.   _____

18. Logical operators are often used to extract information from relational databases. Suppose you were managing a database of demographic information including the proposition p "is a student" and q "owns a car." How would you search this database for students who do not own cars?   _____

19. Programmers can often simplify code by substituting logically equivalent expressions. Create truth tables to show that the expressions $\neg(p \lor q)$ and $\neg p \land \neg q$ are equivalent.   _____

20. Logical expressions, when built into computer hardware, are referred to as gates. Create a truth table for a *NAND* gate, which combines the *NOT* and *AND* gates in the form $\neg(p \land q)$.   _____

Unit 34 LOGICAL NOTATION

BASIC PRINCIPLES

The various fields of study such as mathematics, computer science, and electronics often use different symbols to express logical operations. The following table lists some of the symbols that are commonly used.

	AND		OR		NOT	
Set Theory	$A \cap B$		$A \cup B$		A^C	A'
Logic	$A \wedge B$		$A \vee B$		$\neg A$	A'
Programming	A&B	A&&B	A|B	A||B	~A	!A
Electronics	AB	$A \cdot B$	$A + B$		$\overline{A}$	

Example: How might a computer electronics technician express the logical operation $\neg(A \vee B)$?

$$\overline{A + B}$$

Example: How might a computer programmer express the same operation?

!(A\|\|B)

REVIEW PROBLEMS

Convert the following Boolean (logical) expressions into logic notation:

1. !(*a*&&*b*) _____

2. $x \cdot (z + y)$ _____

3. *yy'* _____

4. *p*\|\|(!*q*&&*r*) _____

5. *pq'r* _____

Convert the following Boolean expressions into programming notation:

6. $x \wedge (y \vee z)$ _____

7. $ab + \bar{c}$ _____

8. $(a \vee b)(a' \vee b)$ _____

9. $x'\,y \vee xyz'$ _____

10. $\overline{(a + b)c}$ _____

Convert the following Boolean expressions into electronics notation:

11. $(x \vee y)'$ _____

12. $\neg p \wedge \neg q$ _____

13. $a \wedge (a' \vee b)$ _____

14. $p \| !p = 1$ _____

15. $x \&\& (y \| x) = (x \&\& y) \| (x \&\& z)$ _____

Here are some practical problems that will apply what you have learned.

16. A programmer needs to simulate the function of an electronic circuit that can be described with the expression $ab'c \vee ac'$. Convert this into programming notation.   _____

17. A distributive property of Boolean logic states that $x \wedge (y \vee z) = (x \wedge y) \vee (x \wedge z)$. Given that this is true, how could you simplify a function in a program that looks like $f = (x \&\& y) \| (x \&\& z)$? **Hint:** Before simplifying, translate into a common form of logic.   _____

18. A computer electronics technician wants to test a NAND gate ($\overline{AB}$), but first she wishes to convert it into logical notation. What would that look like?   _____

19. A database manager merges three fields according to the expression $(A \cup B') \cap C$. What is the equivalent logical expression?   _____

20. You are debugging the following C++ program. What is the truth value of
    walkToday?   _____

```
//It is Sunny
bool itIsSunny = true;
//It is cold outside
bool itIsWarm = false;
//Nice day for walk?
bool walkToday = itIsSunny && itIsWarm;
```

Appendix

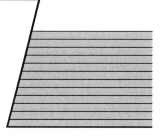

SECTION I

USING A CALCULATOR

Typical Key Symbols and Functions on a Scientific Calculator

+	Addition
−	Subtraction
×	Multiplication
÷	Division
=	Equals
≈	Nearly equals
+/−	Change sign
( )	Parentheses
EE	Scientific notation
Eng	Engineering notation
x^2	Square
y^x	y to the power of x
$\sqrt{x}$	Square root
$1/x$	Reciprocal
%	Percent
log	Logarithm
ln x	Natural log
Hex	Hexadecimal
Bin	Binary
Dec	Decimal

SECTION II

METRIC CONVERSION

Length

1 inch (in)	≈	25.4 millimeters (mm)
1 inch (in)	≈	2.54 centimeters (cm)
1 foot (ft)	≈	0.305 meters (m)
1 yard (yd)	≈	0.914 meters (m)
1 mile (mi)	≈	1.60 kilometers (km)
1 millimeter (mm)	≈	0.0394 inches (in)
1 centimeter (cm)	≈	0.394 inches (in)
1 meter (m)	≈	3.28 feet (ft)
1 kilometer (km)	≈	0.621 miles (mi)

Area

1 square inch (sq in)	≈	645 square millimeters (mm^2)
1 square inch (sq in)	≈	6.45 square centimeters (cm^2)
1 square foot (sq ft)	≈	0.0929 square meters (m^2)
1 square yard (sq yd)	≈	0.836 square meters (m^2)
1 square millimeter (mm^2)	≈	0.00155 square inches (sq in)
1 square centimeter (cm^2)	≈	0.155 square inches (sq in)
1 square meter (m^2)	≈	10.8 square feet (sq ft)

Volume

1 cubic inch (cu in)	≈	16.4 cubic centimeters (cm³)
1 cubic foot (cu ft)	≈	0.0283 cubic meters (m³)
1 cubic yard (cu yd)	≈	0.765 cubic meters (m³)
1 gallon (gal)	≈	3.79 liters (L)
1 quart (qt)	≈	0.946 liters (L)
1 fluid ounce (oz)	≈	29.6 cubic centimeters (cm³)
1 cubic centimeter (cm³)	≈	0.0610 cubic inches (cu in)
1 cubic centimeter (cm³)	≈	0.0338 fluid ounces (oz)
1 liter (l)	≈	1.06 quarts (qt)
1 cubic meter (m³)	≈	1.31 cubic yards (cu yd)

Mass

1 pound (lb)	≈	0.454 kilograms (kg)
1 ounce (oz)	≈	28.3 grams (g)
1 gram (g)	≈	0.0353 ounces (oz)
1 kilogram (kg)	≈	2.20 pounds (lbs)

SECTION III

ASCII TABLE

DEC	HEX	Symbol	DEC	HEX	Symbol	DEC	HEX	Symbol
32	20	Space	64	40	@	96	60	`
33	21	!	65	41	A	97	61	a
34	22	"	66	42	B	98	62	b
35	23	#	67	43	C	99	63	c
36	24	$	68	44	D	100	64	d
37	25	%	69	45	E	101	65	e
38	26	&	70	46	F	102	66	f
39	27	'	71	47	G	103	67	g
40	28	(	72	48	H	104	68	h
41	29	)	73	49	I	105	69	i
42	2A	*	74	4A	J	106	6A	j
43	2B	+	75	4B	K	107	6B	k
44	2C	,	76	4C	L	108	6C	l
45	2D	–	77	4D	M	109	6D	m
46	2E	.	78	4E	N	110	6E	n
47	2F	/	79	4F	O	111	6F	o
48	30	0	80	50	P	112	70	p
49	31	1	81	51	Q	113	71	q
50	32	2	82	52	R	114	72	r
51	33	3	83	53	S	115	73	s
52	34	4	84	54	T	116	74	t
53	35	5	85	55	U	117	75	u
54	36	6	86	56	V	118	76	v
55	37	7	87	57	W	119	77	w
56	38	8	88	58	X	120	78	x
57	39	9	89	59	Y	121	79	y
58	3A	:	90	5A	Z	122	7A	z
59	3B	;	91	5B	[	123	7B	{
60	3C	<	92	5C	\	124	7C		
61	3D	=	93	5D	]	125	7D	}
62	3E	>	94	5E	^	126	7E	~
63	3F	?	95	5F	_	127	7F	[Del]

SECTION IV

COMMON FORMULAS

Perimeter

Square
$P = 4s$

P = perimeter
s = side

Rectangle
$P = 2l + 2w$

P = perimeter
l = length
w = width

Triangle
$P = a + b + c$

P = perimeter
a = first side
b = second side
c = third side

Circle
$C = 2\pi r = \pi D$

C = circumference
π = 3.1416 (approximate)
r = radius
D = diameter

Area

Square
$A = s^2$

A = area
s = side

Rectangle
$A = lw$

A = area
l = length
w = width

Triangle
$A = \frac{1}{2} bh$

A = area
b = base
h = height

Circle
$A = \pi r^2$

A = area
π = 3.1416 (approximate)
r = radius

Ellipse
$A = \pi ab$

A = area
π = 3.1416 (approximate)
a = semimajor axes
b = semiminor axes

Volume

Rectangular Solid V = volume
$V = lwh$ l = length
                                  w = width
                                  h = height

Cylinder V = volume
$V = \pi r^2 h$ π = 3.1416 (approximate)
                                  r = radius
                                  h = height

Glossary

802.11 — A wireless fidelity (Wi-Fi) standard developed by the IEEE (Institute for Electrical and Electronics Engineers) for transmitting data over a wireless network.

AGP (Accelerated Graphics Port) — A graphics card expansion port attached to a computer's motherboard to enhance performance of video and animation.

Amp (ampere) — A unit of electric current designated with the symbol A

ASCII (American Standard Code for Information Interchange) — The universal standard for the numerical codes computers use to represent all upper- and lowercase letters, numbers, and punctuation.

ATA (Advanced Technology Attachment) — A standard for attaching storage devices to a computer. ATA drives often are also referred to using the term IDE (integrated device electronics).

ATX — A form factor (the shape and layout of computer components) that specifies the size, shape, and orientation of motherboards, cases, and power supplies. (Applies also to Mini-ATX and Micro-ATX.)

Backside Bus — The bus that transfers data to and from a computer's secondary cache. See also Frontside Bus and Bus.

BIOS (Basic Input/Output System) — Firmware code (i.e., software that is a permanent part of the hardware) controlling a computer's basic functions at start-up.

Bit (Binary Digit) — A 1 or 0 used as the basic unit of digital data storage.

Bluetooth — A standard for short-range wireless communication, often used for peripheral equipment such as wireless computer mice or speakers.

Boot — The process of starting a computer and loading the operating system.

Bus — Any of a number of sets of wires inside a computer that allow data to be passed back and forth to different parts of the machine. Each bus has a certain size (such as 32-bit or 64-bit), which determines how much data can travel across the bus at one time. Buses also have a certain speed (measured in MHz), which determines how fast the data can travel.

Byte — A collection of 8 bits.

Cat 5 (also Cat 5e, Cat 6) — Cat stands for "category" and is a specification for unshielded twisted pair cable often used in local area networks.

Cluster — A group of sectors forming an allocation unit that is used to organize and identify files on the disk; the smallest logical division of a hard drive.

CMOS (Complementary Metal Oxide Semiconductor) — A battery that powers the CMOS chip, which stores basic configuration information about the computer.

Coaxial Cable — A type of networking cable used to transmit telephone and television signals of high frequency.

Color Depth — The number of colors displayed on a computer screen.

COM Port (Communications Port) — A serial port for computer peripherals (such as mice); has been largely replaced by the USB port.

CPU (Central Processing Unit) — The main microprocessor in a computer.

CRT (Cathode Ray Tube) — The technology used to create conventional televisions and computer monitors; has been largely replaced by flat screen displays.

Differential Backup — A method of backing up only the files that have changed since the last full backup.

DIMM (Dual In-line Memory Module) — A kind of memory chip used in personal computers.

DSL (Digital Subscriber Line) — A type of high-speed Internet connection that sends digital data through a telephone line.

Ethernet — A technology used for local area networks.

Expansion Bus — Any of a number of devices that allow data to move to and from expansion cards, including video cards and other I/O devices. See also Bus, AGP, and PCI.

FAT (File Allocation Table) — A file system that divides up the contents of a drive into clusters.

Firewire — A high-speed serial interface similar to USB; also referred to by its technical name, IEEE 1394, since being standardized by the Institute of Electrical and Electronics Engineers. Firewire and IEEE 1394 refer to the same technology.

Frame Buffer — The memory area on a video display used to store color values for every pixel.

Frontside Bus (System Bus) — The computer's primary high-speed connection between the CPU and the rest of the components on a motherboard.

Heat-Sink Paste — A thermal grease compound used between the CPU and the heat sink to ensure good thermal conductivity.

Hertz (Hz) — A unit for frequency, expressed in cycles per seconds.

Host — A computer or device such as a printer on a network.

Hub — A generic term for a network device that retransmits signals; it usually indicates a repeater or level 1 switch.

IDE (Integrated Drive Electronics) — See ATA.

Incremental Backup — A method of backing up only the files that have changed since the last backup of any kind.

I/O Address — Numbers that are assigned to devices that need to communicate with the CPU. Also called port addresses.

IP Address — Numbers used to identify computers and devices on a network that uses the TCP/IP protocol.

IRQ (Interrupt Request) — A signal sent from a device (such as a modem or keyboard) to the CPU. Each device has its own IRQ address to enable the CPU to sequence requests efficiently.

ISA (Industry Standard Architecture) — An old type of bus used for attaching cards such as modems directly to a motherboard. Largely replaced by AGP and PCI expansion buses.

ISDN (Integrated Services Digital Network) — A telephone system that transmits both voice and data digitally.

KVM — Stands for "Keyboard Video Mouse" and refers to a device that allows several computers to be controlled from the same keyboard, monitor, and mouse.

LAN (Local Area Network) — A network created with cables that connects computers located within a small area such as an office.

LCD (Liquid Crystal Display) — A type of flat-panel display for televisions and computer monitors.

Linux — An open-source operating system for personal computers.

MAC (Media Access Control) Address — A physical address assigned to all devices on a TCP/IP network.

Motherboard — The main circuit board inside a computer; also called the main board or the system board.

Multimode Fiber — A type of short-range fiber optic cable often used within a local area network.

Nibble — A collection of four bits or half of a byte.

OC (Optical Carrier) — A fiber optic network specification.

Packet — A section of data sent on a network; also called a data packet.

Parity — In computing, a form of error checking that determines if data has been lost or overwritten during transmission.

Patch Cable — A cable used to connect a computer to a network.

PC Card — A peripheral interface often used in portable computers; a shortened version of Personal Computer Memory Card International Association (PCMCIA) card.

PCI (Peripheral Component Interconnect) — A bus used for attaching devices directly to a computer motherboard; replaces the older ISA bus.

PCIe or **PCI Express** — A high-speed interface meant to replace the PCI and AGP expansion buses.

PCMCIA Card — See PC Card.

Ping — Stands for "Packet Internet Groper" and is a command used to verify a connection between computers on a network.

Pixel — Short for "picture element" and refers to a single point on a monitor.

POST (Power-On Self-Test) Card — An adapter card used in computer troubleshooting during the boot up process.

Process — A program combined with its resources running on a computer.

Protected Mode — A 32-bit CPU mode that supports features such as multitasking and virtual memory; also referred to as 32-bit mode.

Punch Down Block — A type of connector often used in networking; it allows wires to be connected quickly.

RAID (Redundant Array of Independent Disks) — A standard for combining several hard drives to increase drive performance or reliability.

RAM (Random Access Memory) — A kind of temporary storage that holds programs and data while the computer is running; allow programs to run faster than accessing them from the hard drive.

Real Mode — A 16-bit CPU mode that does not support features such as multitasking or virtual memory; also called 16-bit mode.

Registry — A Windows database used to store configuration information.

Repeater — A network device that retransmits signals on a local area network (see also Hub).

Resolution — A measure of the number of pixels on a computer display screen.

RFID (Radio-Frequency Identification) **Tag** — The transponder device in an RFID system (which also consists of an antenna and a transceiver); used for remote identification through radio waves.

RGB — Stands for "red green blue," the three primary colors that, when combined in different intensities, create all of the colors in computer and television displays

RJ-45 — A type of connector used with unshielded twisted pair cables.

Router — A network device that connects different networks.

SATA (Serial ATA) — A high-speed serial interface for hard drives.

SCSI (Small Computer System Interface) — An interface primarily for attaching high-speed peripheral devices such as hard drives or scanners. SCSI is pronounced, "skuzzy."

SCSI ID — A unique number assigned to each SCSI device on a computer.

Sector — The smallest physical unit that can be accessed on a hard disk.

Singlemode Fiber — A type of high-bandwidth fiber optic cable often used for long-distance communication.

Site License — A license granted by the developer to install a piece of software on more than one computer.

SQL (Structured Query Language) — A computer language created for accessing and working with databases; is pronounced either "S-Q-L" or "sequel."

Subnet Mask — A number that defines a range of IP addresses that can be used in a network.

Switch — A network device used to reduce network traffic by segmenting the network into separate collision domains. A switch is more capable than a hub, but less so than a router.

T Carrier — A specification of network bandwidth; T1 and T3 refer to high-speed data transfer systems that have very large bandwidths.

TCP/IP protocol — A common set of rules governing the communication between Internet devices.

Terminator — A device used at the end of a network cable to reduce interference.

Thread — An individual task being run on a CPU.

Throughput — The amount of data that can be sent through a network cable.

TLA — Stands for "three letter acronym"; a self-referential abbreviation.

Traceroute — A command used to determine the path taken by packets across a TCP/IP network.

Track — A concentric ring around a hard drive platter.

UPS (Uninterruptible Power Supply) — A device used to supply backup power.

USB (Universal Serial Bus) — A serial interface used for peripherals such as computer mice and printers.

Voltage — An electrical force, expressed in volts, and the symbols for which are V and E.

Watt — A unit of electrical power whose symbol is W.

Wave File — A type of uncompressed audio file format, whose file name extension is WAV (.wav).

Answers to Odd-Numbered Problems

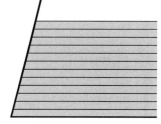

SECTION 1   WHOLE NUMBERS

Unit 1   Addition of Whole Numbers

1. 200 GB
3. 44 threads
5. 60 watts
7. 71 processes

9. 73 computers
11. 653 computers
13. 477 calls
15. 120 minutes or 2 hours

17. 8,505 MB
19. Yes

Unit 2   Subtraction of Whole Numbers

1. $88
3. 39 minutes
5. $28
7. 384 MB

9. 158 were not returned
11. 50 seconds
13. the symbol "5"
15. 27 feet

17. 1 port
19. No, there will still be 112 GB of files on the drive

Unit 3   Multiplication of Whole Numbers

1. 768 MB
3. 2,132 MHz or approximately 2.1 GHz
5. 307,200 pixels

7. 1,024 bits
9. 360 watts
11. 3,600 frames
13. $675

15. 540,000 dots
17. 1,472 Kbps
19. 48 bits

Unit 4   Division of Whole Numbers

1. 54 Mbps
3. 14 boxes
5. 128 KB each
7. 48 seconds

9. 12 Kbps
11. 25 seconds
13. No, it will take you 40 minutes.

15. 12 blocks
17. 28 connections per office
19. 95 GB

Unit 5   Combined Operations with Whole Numbers

1. 141,363 bytes
3. 840 minutes
5. 1,530 MB
7. 2,400 Mbps
9. 12 hours
11. 2,544 cents or $25.44
13. 65,520 B
15. 180 cm
17. 59 problems solved
19. 92,160 bytes per second

SECTION 2   COMMON FRACTIONS

Unit 6   Addition of Common Fractions

1. 10¼ feet
3. 9⁵⁄₁₆ inches
5. 190 connectors
7. 14 hops
9. 58⁹⁄₁₀ meters
11. ²³⁄₄₀ of the code
13. 8⅛ inches
15. 4⅓ inches
17. ⅝ inch
19. 42½ hours

Unit 7   Subtraction of Common Fractions

1. ⁵⁄₁₂ more ink
3. Yes
5. ¼ of the packets
7. 5⁵⁄₁₂ inches
9. ¹³⁄₁₅
11. 9¼ feet
13. ⁵⁄₃₂ inch
15. ⁴⁷⁄₆₄ represents other characters
17. 110 GB
19. 5 hours

Unit 8   Multiplication of Common Fractions

1. 24⅔ hours
3. 24 bits
5. ⅜
7. ½ of the computers will be affected
9. ¹⁄₃₆ GB
11. 21 unsecured networks
13. 8¾ inches
15. No
17. ⅑ of the packets
19. $1,240

Unit 9   Division of Common Fractions

1. 15 grams of heat-sink paste
3. 28 cables
5. $20 per hour
7. 4,725 computer systems
9. 7,200 rpm
11. 128 watts
13. 800 MHz
15. 141 MB
17. ⁷⁄₁₅ TB
19. 96 bits

Unit 10   Combined Operations with Common Fractions

1. 3,200 MB
3. No
5. 7 connections
7. 1¼ hours

9. 1,200 GB
11. 20 GB
13. 4 KVM switches
15. 3 programmers

17. $57
19. 1/18 of its time

SECTION 3   DECIMAL FRACTIONS

Unit 11   Addition of Decimal Fractions

1. 45.6 GB
3. $205.52
5. 38.8 inches
7. 215.8 MB

9. 22.9 days
11. 72.5 meters
13. 13.2 GB
15. 77.7 watts

17. 101.7 cm
19. 341.05 MB

Unit 12   Subtraction of Decimal Fractions

1. 0.6 volts
3. 6.53 GB
5. 2.4 volts
7. No

9. 0.7 volts
11. 33.9 meters
13. $4.40 per hour
15. 471.60 MB

17. 1.9 GBps
19. 12.4 seconds

Unit 13   Multiplication of Decimal Fractions

1. 3,600 MHz
3. 9.4 GB
5. 216 watts
7. 8.4 MBps

9. $273.75
11. $43.70
13. 201.3 square inches or 201.3 in^2

15. 495.6 MB
17. 450 watts
19. 22.4 GB

Unit 14   Division of Decimal Fractions

1. 6.7 milliliters (rounds up to 7 milliliters)
3. 1.9 times
5. 42

7. $0.06
9. 1.75 inches
11. 0.25 GBps or 250 MBps
13. 189.56 GB

15. 15 minutes
17. 12.43 MB
19. $1.18 per GB

Unit 15   Combined Operations with Decimal Fractions

1. 8.33 MBps
3. 9 GBps
5. $107.59
7. 9 CPUs

9. 514.81 watts
11. $645.58
13. 130 minutes
15. 24.5 watts

17. 15.75 inches
19. $0.26

SECTION 4   STATISTICS AND ESTIMATES

Unit 16   Percent and Percentage

1. $39
3. 4.4%
5. 20%
7. 33%

9. 360 watts
11. 56%
13. 32.0%
15. 76%

17. 14.1%
19. 0.0075

Unit 17   Interest and Discounts

1. $400
3. $2,916.67
5. $2,093.33
7. $1,651.42

9. 14%
11. $119.99
13. $52.46
15. $180.00

17. A 30% discount
19. $6,318.75

Unit 18   Averages and Estimates

1. 26 Kbps
3. $227.20
5. 6.9 pages per month
7. 34.7 degrees

9. 36.1 hours
11. 13.5 hours
13. 900 minutes or 15 hours
15. $191.80

17. 83.3
19. 11%

SECTION 5   UNITS AND NOTATION

Unit 19   Exponents

1. $2^1 = 2$, $2^2 = 4$, $2^3 = 8$,
    $2^4 = 16$, $2^5 = 32$, $2^6 = 64$,
    $2^7 = 128$, $2^8 = 256$,
    $2^9 = 512$, $2^{10} = 1,024$
3. 65,536
5. 1,048,576

7. 78.5 square centimeters
    or 78.5 cm^2
9. 11,304 square mm or
    11,304 mm^2
11. 134,217,728 KB
13. 2,048

15. 144.70 cubic inches or
    144.70 in^3
17. 65,536
19. 15,625,000

Unit 20   Scientific and Engineering Notation

1. 3.2×10^9
3. 8.192×10^3
5. 1.25×10^{-2}
7. 150×10^6

9. 8×10^{-6} meters
11. 395,000,000 hosts
13. 1,066,000,000 Hz
15. 0.000000075 seconds

17. 0.000001550 meters
19. 700,000,000 bytes

Unit 21   Significant Digits and Rounding

1. 3
3. 2
5. 3
7. 400 feet or 4.0×10^2 feet
   (2 significant figures)

9. 42 MB
11. 2,400 MHz
13. 319 square inches

15. 150 fps
17. 560 GB
19. $100

Unit 22   Equivalent Units

1. 2.896 volts
3. 12.463287 MB
5. 4,500 Kpbs
7. 10,000,000 milliseconds

9. 45,000,000 millimeters
11. 4,700 MB
13. 0.1 mF
15. 4,000,000 megaflops

17. 54,800 meters
19. 0.001 microseconds

SECTION 6   FORMULAS AND SYMBOLS

Unit 23   Functions and Formulas

1. 10
3. 15.7 feet
5. 24 watts
7. $5.88

9. 3.2 MBps
11. 900 KB
13. 120 mm
15. 5 kg

17. $lbs = \dfrac{kg}{0.454}$

19. $charge = (hours \times 25) + 40$

Unit 24   Exponential and Logarithmic Functions

1. 19.625 square inches
3. 6.9 volts
5. 1,728 cubic inches
7. 33.5 cubic feet

9. 78.125 foot-candles
11. 59.38 square inches
13. 31 inches
15. 31 meters

17. 2
19. 86.4

Unit 25   Rearranging Formulas

1. $x = \dfrac{10}{2} = 5$

3. $b = 45 - 16 = 29$

5. $b = 2$

7. $n = 0$

9. $x = 626$

11. $d = \dfrac{c}{\pi}$

13. $x = \dfrac{y - b}{m}$

15. $b = y - mx$

17. $volts = \dfrac{watts}{amps}$

19. $x = \sqrt{z^2 - y^2}$

SECTION 7   NUMBERING SYSTEMS

Unit 26   Binary Representation

1. 170

3. 000, 001, 010, 011, 100, 101, 110, 111

5. 00001010.00001101. 01101110.11010001

7. 00000000 to 01111111 = 0 to 127

9. Yes

11. 255.252.0.0

13. 1011 = 11 in decimal

15. 33

17. 63 = 111111 = 6 bits

19. 7 = 00000111, Yes

Unit 27   Binary Addition and Multiplication

1. 111

3. 11100

5. 11001100

7. 1101100011

9. 10000

11. 5 bits

13. 15

15. 16,777,216

17. 65,536

19. 128

Unit 28   Hexadecimal and Octal Numbering

1. F

3. 1,016 to 1,023

5. 5h

7. Yes

9. 8 bits

11. $32 \times 1024 = 32768 =$ 8000h, 8000h + C0000h = C8000h

13. FFFFFh – F0000h = FFFFh = 65,535 bytes

15. 12

17. 7

19. 2000

Unit 29   BCD and ASCII Encoding

1. 2
3. May
5. 127 = 0001 0010 0111
     = 12 bits
7. 0110 0100

9. S
11. 48 65 6C 6C 6F
13. 01001000 01101001
15. 32

17. 00001100
19. 66 69 72 73 74 2E 6C 61
      73 74 40 77 6F 72 6B 2E
      63 6F 6D

SECTION 8   SETS

Unit 30   Set Theory

1. U = {Windows XP,
    Windows Vista, Linux,
    MAC OS}
3. U = {00, 01, 10, 11}
5. U = {00h, 01h, 02h, . . . ,
    7Dh ,7Eh, 7Fh}

7. U = {Windows operating
    systems created before
    1960}
9. Yes
11. Yes

13. Yes
15. Yes
17. 4
19. 4

Unit 31   Operators

1. 3
3. 16
5. $A \cap B$ = {2, 6}
7. F' = {Carlos, Tom, Ahmed}

9. 2
11. $A \cup B$ = {000, 001, 011,
     100, 101, 110}
13. $(A \cup B)'$ = {010, 111}

15. 5
17. 4
19. $A \cup (B \cap C)$ = {2, 4, 6}

Unit 32   Venn Diagrams

1.

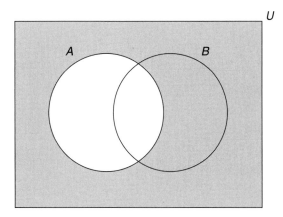

3.

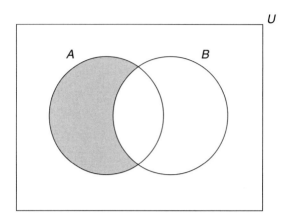

5.

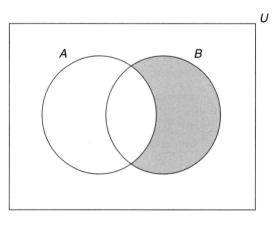

7.

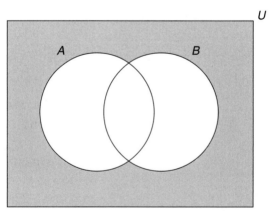

9.

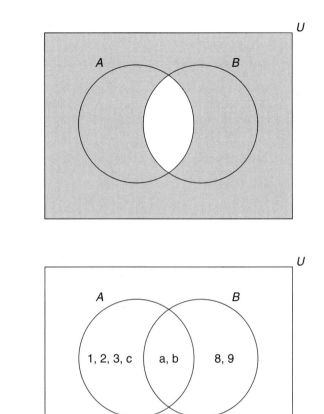

11.

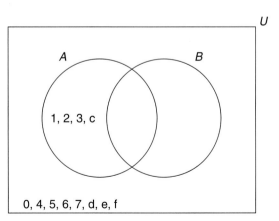

13.

15.

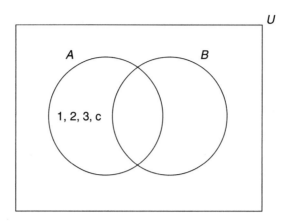

17.

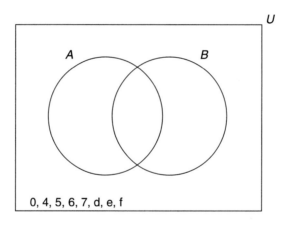

19.

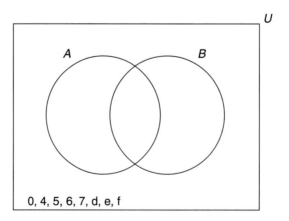

SECTION 9   LOGIC

Unit 33   Truth Tables

1. Yes
3. No
5. Yes
7. $p \wedge \neg q$

p	q	$p \wedge \neg q$
1	1	0
1	0	1
0	1	0
0	0	0

9. $\neg p \wedge \neg q$

p	q	$\neg p \wedge \neg q$
1	1	0
1	0	0
0	1	0
0	0	1

11. True
13. False
15. True
17. Carolina AND (North OR South)
19.

p	q	$\neg(p \vee q)$	$\neg p \vee \neg q$
1	1	0	0
1	0	0	0
0	1	0	0
0	0	1	1

Unit 34   Logical Notation

1. $\neg(A \wedge B)$
3. $y \wedge \neg y$
5. $p \wedge \neg q \wedge r$
7. $a \&\& b \,\|\, !c$
9. $!x \&\& y \,\|\, x \&\& y \&\& !z$
11. $\overline{x + y}$
13. $a(\overline{a} + b)$
15. $x(y + x) = xy + xz$
17. $x \&\& (y \,\|\, z)$
19. $(a \vee \neg b) \wedge c$